THINKING ABOUT STATISTICS

Simply stated, this book bridges the gap between statistics and philosophy. It does this by delineating the conceptual cores of various statistical methodologies (Bayesian/frequentist statistics, model selection, machine learning, causal inference, etc.) and drawing out their philosophical implications. Portraying statistical inference as an epistemic endeavor to justify hypotheses about a probabilistic model of a given empirical problem, the book explains the role of ontological, semantic, and epistemological assumptions that make such inductive inference possible. From this perspective, various statistical methodologies are characterized by their epistemological nature: Bayesian statistics by internalist epistemology, classical statistics by externalist epistemology, model selection by pragmatist epistemology, and deep learning by virtue epistemology.

Another highlight of the book is its analysis of the ontological assumptions that underpin statistical reasoning, such as the uniformity of nature, natural kinds, real patterns, possible worlds, causal structures, etc. Moreover, recent developments in deep learning indicate that machines are carving out their own "ontology" (representations) from data, and better understanding this—a key objective of the book—is crucial for improving these machines' performance and intelligibility.

Key Features

- Without assuming any prior knowledge of statistics, discusses philosophical aspects of traditional as well as cutting-edge statistical methodologies.
- Draws parallels between various methods of statistics and philosophical epistemology, revealing previously ignored connections between the two disciplines.
- Written for students, researchers, and professionals in a wide range of fields, including philosophy, biology, medicine, statistics and other social sciences, and business.
- Originally published in Japanese with widespread success, has been translated into English by the author.

Jun Otsuka is Associate Professor of Philosophy at Kyoto University and a visiting researcher at the RIKEN Center for Advanced Intelligence Project in Saitama, Japan. He is the author of *The Role of Mathematics in Evolutionary Theory* (Cambridge UP, 2019).

THINKING ABOUT STATISTICS

The Philosophical Foundations

Jun Otsuka

Routledge
Taylor & Francis Group
NEW YORK AND LONDON

Designed cover image: © Jorg Greuel/Getty Images

First published in English 2023
by Routledge
605 Third Avenue, New York, NY 10158

and by Routledge
4 Park Square, Milton Park, Abingdon, Oxon, OX14 4RN

Routledge is an imprint of the Taylor & Francis Group, an informa business

© 2023 Jun Otsuka

The right of Jun Otsuka to be identified as author of this work has been asserted in accordance with sections 77 and 78 of the Copyright, Designs and Patents Act 1988.

All rights reserved. No part of this book may be reprinted or reproduced or utilised in any form or by any electronic, mechanical, or other means, now known or hereafter invented, including photocopying and recording, or in any information storage or retrieval system, without permission in writing from the publishers.

Trademark notice: Product or corporate names may be trademarks or registered trademarks, and are used only for identification and explanation without intent to infringe.

Originally published in Japanese by the University of Nagoya Press, Japan.

ISBN: 978-1-032-33310-6 (hbk)
ISBN: 978-1-032-32610-8 (pbk)
ISBN: 978-1-003-31906-1 (ebk)

DOI: 10.4324/9781003319061

Typeset in Bembo
by Apex CoVantage, LLC

For my parents, Yuzuru and Kuniko Otsuka

CONTENTS

Preface to the English Edition   x

Introduction   1
   What Is This Book About?   1
   The Structure of the Book   4

1   The Paradigm of Modern Statistics   10
   1.1   Descriptive Statistics   10
      1.1.1   Sample Statistics   11
      1.1.2   Descriptive Statistics as "Economy of Thought"   13
      1.1.3   Empiricism, Positivism, and the Problem of Induction   16
   1.2   Inferential Statistics   17
      1.2.1   Probability Models   18
      1.2.2   Random Variables and Probability Distributions   21
      1.2.3   Statistical Models   27
      1.2.4   The Worldview of Inferential Statistics and "Probabilistic Kinds"   34
   Further Reading   38

2   Bayesian Statistics   41
   2.1   The Semantics of Bayesian Statistics   41
   2.2   Bayesian Inference   46
      2.2.1   Confirmation and Disconfirmation of Hypotheses   47
      2.2.2   Infinite Hypotheses   48
      2.2.3   Predictions   50

2.3 Philosophy of Bayesian Statistics 51
    2.3.1 Bayesian Statistics as Inductive Logic 51
    2.3.2 Bayesian Statistics as Internalist Epistemology 53
    2.3.3 Problems with Internalist Epistemology 57
    2.3.4 Summary: Epistemological Implications of Bayesian Statistics 70
Further Reading 72

3   Classical Statistics   74
  3.1 Frequentist Semantics 75
  3.2 Theories of Testing 79
    3.2.1 Falsification of Stochastic Hypotheses 79
    3.2.2 The Logic of Statistical Testing 80
    3.2.3 Constructing a Test 81
    3.2.4 Sample Size 84
  3.3 Philosophy of Classical Statistics 85
    3.3.1 Testing as Inductive Behavior 85
    3.3.2 Classical Statistics as Externalist Epistemology 87
    3.3.3 Epistemic Problems of Frequentism 94
    3.3.4 Summary: Beyond the Bayesian vs. Frequentist War 104
  Further Reading 106

4   Model Selection and Machine Learning   109
  4.1 The Maximum Likelihood Method and Model Fitting 109
  4.2 Model Selection 112
    4.2.1 Regression Models and the Motivation for Model Selection 112
    4.2.2 A Model's Likelihood and Overfitting 114
    4.2.3 Akaike Information Criterion 115
    4.2.4 Philosophical Implications of AIC 118
  4.3 Deep Learning 124
    4.3.1 The Structure of Deep Neural Networks 124
    4.3.2 Training Neural Networks 126
  4.4 Philosophical Implications of Deep Learning 129
    4.4.1 Statistics as Pragmatist Epistemology 129
    4.4.2 The Epistemic Virtue of Machines 131
    4.4.3 Philosophical Implications of Deep Learning 136
  Further Reading 142

5   Causal Inference   144
  5.1 The Regularity Theory and Regression Analysis 145
  5.2 The Counterfactual Approach 149
    5.2.1 The Semantics of the Counterfactual Theory 149
    5.2.2 The Epistemology of Counterfactual Causation 151

    5.3 Structural Causal Models 158
        5.3.1 Causal Graphs 159
        5.3.2 Interventions and Back-Door Criteria 162
        5.3.3 Causal Discovery 165
    5.4 Philosophical Implications of Statistical Causal Inference 167
    Further Reading 171

6   The Ontology, Semantics, and Epistemology of Statistics   172
    6.1 The Ontology of Statistics 172
    6.2 The Semantics of Statistics 176
    6.3 The Epistemology of Statistics 179
    6.4 In Lieu of a Conclusion 181

Bibliography   *183*
Index   *189*

PREFACE TO THE ENGLISH EDITION

This book is the English edition of a book originally published in Japanese under the title *Tokeigaku wo Tetsugaku Suru (Philosophizing About Statistics)*, by the University of Nagoya Press. Instead of composing a word-to-word translation, I took this occasion to revise the whole book by incorporating feedback to the original edition and adding some new paragraphs, so it might be more appropriate to call it a rewrite rather than a translation. I also replaced some references and book guides at the end of the chapters with those more accessible to English readers.

Translating your own writing into a language of which you don't have a perfect command is a painful experience. It was made possible only by close collaboration with Jimmy Aames, who went through every sentence and checked not only my English but also the content. Needless to say, however, I am responsible for any errors that may remain. I also owe a great debt to Yukito Iba, Yoichi Matsuzaka, Yusaku Ohkubo, Yusuke Ono, Kentaro Shimatani, Shohei Shimizu, and Takeshi Tejima for their comments on the original Japanese manuscript and book, and Donald Gillies, Clark Glymour, Samuel Mortimer, and an anonymous reviewer for Routledge for their feedback on the English manuscript, all of which led to many improvements.

This book, like all of my other works, was made possible by the support of my mentors, teachers, and friends, including but not limited to: Steen Andersson, Yasuo Deguchi, Naoya Fujikawa, Takehiko Hayashi, Chunfeng Huang, Kunitake Ito, Manabu Kuroki, Lisa Lloyd, Guilherme Rocha, Robert Rose, and Tomohiro Shinozaki. I am also grateful to Kenji Kodate and his colleagues at the University of Nagoya Press, and Andrew Beck and his colleagues at Routledge, for turning the manuscript into books.

Finally and most of all, I am grateful to my family, Akiko and Midori Otsuka, for their moral support during my writing effectively two books in a row in the midst of the global upheaval caused by the COVID-19 pandemic.

INTRODUCTION

What Is This Book About?

This book explores the intersection between statistics and philosophy, with the aim of *introducing philosophy to data scientists and data science to philosophers*. By "data science" I am not referring to specific disciplines such as statistics or machine learning research; rather, I am using the term to encompass all scientific as well as practical activities that rely on quantitative data to make inferences and judgments. But why would such a practical science have anything to do with philosophy, often caricatured as empty armchair speculation? Statistics is usually regarded as a rigid system of inferences based on rigorous mathematics, with no room for vague and imprecise philosophical ideologies. A philosophically minded person, on the other hand, might dismiss statistics as merely a practical tool that is utterly useless in tackling deep and ineffable philosophical mysteries.

The primary aim of this book is to dispel these kinds of misconceptions. Statistics today enjoys a privileged role as *the* method of deriving scientific conclusions from observed data. For better or worse, in most popular and scientific articles, "scientifically proven" is taken to be synonymous with "approved by a proper statistical procedure." But on what theoretical ground is statistics able to play, or at least expected to play, such a privileged role? The justification of course draws its force from sophisticated mathematical machinery, but how is such a mathematical framework able to justify scientific—that is, empirical—knowledge in the first place? This is a philosophical question *par excellence*, and various statistical methods, implicitly or explicitly, have some philosophical intuitions at their root. These philosophical intuitions are seldom featured in common statistics textbooks, partly because they do not provide any extra tools that readers could use to analyze data they collect for their theses or research projects. However, understanding the philosophical intuitions that lie behind the various statistical methods, such as Bayesian statistics and hypothesis testing, will help one get a grip on their inferential characteristics and make sense of the conclusions obtained from these methods, and thereby become more conscious and responsible about what one is really doing with statistics. Moreover, statistics is by no means a monolith: it comprises a variety of methods and theories, from classical frequentist and Bayesian statistics to the rapidly developing fields of machine learning research, information theory, and causal inference. It goes without saying that the proper application of these techniques demands a firm grasp of their mathematical foundations. At the same time, however, they also involve philosophical intuitions that cannot be reduced to mathematical proofs. These intuitions prescribe, often implicitly, how the world under investigation is structured and how one can make inferences about this world. Or, to use the language of the present book, each statistical method embodies a distinct approach to inductive inference, based on its own characteristic ontology and epistemology. Understanding these ideological backgrounds proves essential in the choice of an appropriate method vis-à-vis a given problem and for the correct interpretation of its results, i.e., in making sound inferences rather than falling back on the routine application of ready-made statistical packages. This is why I believe philosophical thinking, despite its apparent irrelevance, can be useful for data analysis.

But, then, what is the point for a *philosopher* to learn statistics? The standard philosophy curriculum in Japanese and American universities is mostly logic-oriented and does not include much training in statistics, with the possible exception of some basic probability calculus under the name of "inductive logic." Partly because of this, statistics is not in most philosophers' basic toolbox. I find this very unfortunate, because statistics is like an ore vein that is rich in fascinating conceptual problems of all kinds. One of the central problems of philosophy from the era of Socrates is: how can we acquire *episteme*, or true knowledge? This question has shaped the long tradition of epistemology that runs through the modern philosophers Descartes, Hume, and Kant, leading up to today's analytic philosophy. In the course of its history, this question has become entwined with various ontological and/or metaphysical issues such as the assumption of the uniformity of nature, the problem of causality, natural kinds, and possible worlds, to name just a few. As the present book aims to show, statistics is the modern scientific variant of philosophical epistemology that comprises all these themes. That is, statistics is a scientific epistemology that rests upon certain ontological assumptions. Therefore, no one working on epistemological problems today can afford to ignore the impressive development and success of statistics in the past century. Indeed, as we will see, statistics and contemporary epistemology share not only common objectives and interests; there is also a remarkable parallelism in their methodologies. Attending to this parallelism will provide a fruitful perspective for tackling various issues in epistemology and philosophy of science.

Given what has been said thus far, a reader might expect that this book is intended as an introduction to the philosophy of statistics in general. It is not, for two reasons. First, this book does not pretend to introduce the reader to the field of the *philosophy of statistics*, a well-established branch of contemporary philosophy with a wealth of discussions concerning the theoretical ground of inductive inference, interpretations of probability, the everlasting battle between Bayesian and frequentist statistics, and so forth (Bandyopadhyay and Forster 2010). While these are all important and interesting topics, going through them would make a huge volume, and in any case far exceeds the author's capability. Moreover, as these discussions often tend to be highly technical and assume familiarity with both philosophy and statistics, non-specialists may find it difficult to follow or keep motivated. Some of these topics are of course covered in this book, and in the case of others I will point to the relevant literature. But instead of trying to cover all these traditional topics, this book cuts into philosophical issues in statistics with my own approach, which I will explain in a moment. Thus readers should keep in mind that this book is not intended as a textbook-style exposition of the standard views in the philosophy of statistics.

The second reason why this book is not entitled *An Introduction to the Philosophy of Statistics* is that it does not aim to be an "introduction" in the usual sense of the term. The Japanese word for introduction literally means "to enter the gate," with the implication that a reader visits a particular topic and stays there as a guest for a while (imagine visiting a temple) in order to appreciate, experience, and learn its internal atmosphere and architectural art. This book, however, is not a well-mannered tour guide who quietly stays at one topic, either statistics or philosophy. It is indeed a restless traveler, entering the gate of statistics, quickly leaving and entering philosophy from a different gate, only to be found in the living room of statistics at the next moment. At any rate, the goal of this book is not to make the reader proficient in particular statistical tools or philosophical ideas. This does not mean that it presupposes prior familiarity with statistics or philosophy: on the contrary, this book is designed to be as self-contained as possible, providing plain explanations for every statistical technique and philosophical concept at their first appearance (so experts may well want to skip these introductory parts). The aim of these explanations, however, is not to make the reader a master of the techniques and ideas themselves; rather, they are meant to elucidate the conceptual relationships among these techniques and ideas. Throughout this book we will ask questions like: how is a particular statistical issue discussed in the context of philosophy? How does a particular philosophical concept contribute to our understanding of statistical thinking? Through such questions, this book aims to bridge statistics and philosophy and reveal the conceptual parallelism between them. Because of this interdisciplinary character, this book is not entitled "Introduction" and is not intended to be read as such. That is, this book does not pretend to train the reader to become a data scientist or philosopher. Rather, this is a book for border-crossers: it tempts the data analyst to become a little bit of a philosopher, and the philosophy lover to become a little bit of a data scientist.

The Structure of the Book

What kind of topics, then, are covered in this book? This book may be likened to a fabric, woven with philosophy as its warp and statistics as its weft. The philosophy warp consists of three threads: ontology, semantics, and epistemology. *Ontology* is the branch of philosophy that studies the real nature of things existing in the world. Notable historical examples include the Aristotelian theory of the four elements, according to which all subcelestial substances are composed from the basic elements of fire, air, water, and earth; and the mechanical philosophy of the 17th century, which aimed to reduce all physical matter to microscopic particles. But ontology is not monopolized by philosophers. Indeed, every scientific theory makes its own ontological assumptions as to what kinds of things constitute the world that it aims to investigate. The world of classical mechanics, for example, is populated by massive bodies, while a chemist or biologist would claim that atoms and molecules, or genes and cells, also exist according to their worldview. We will not be concerned here with issues such as the adequacy of these ontological claims, or which entities are more "fundamental" and which are "derivative." What I am pointing out is simply the truism that every scientific investigation, insofar as it is an empirical undertaking, must make clear what the study is *about*.

Unlike physics or biology, which have a concrete domain of study, statistics *per se* is not an empirical science and thus may not seem to rely on any explicit assumption about what exists in the world. Nevertheless, it still makes ontological assumptions about the structure of the world in a more abstract way. What are the entities posited by statistics? The first and foremost thing that must exist in statistics is obvious: data. But this is not enough—the true value of statistics, especially its primary component known as inferential statistics, lies in its art of inferring the unobserved from the observed. Such an inference that goes beyond the data at hand is called *induction*. As the 18th-century Scottish philosopher David Hume pointed out, inductive inference relies on what he called the *uniformity of nature* behind the data. Inferential statistics performs predictions and inferences by mathematically modeling this latent uniformity behind the data (Chapter 1). These mathematical models come in various forms, with differing shades of ontological assumptions. Some models assume more "existence" in the world than others, in order to make broader kinds of inferences possible. Although such philosophical assumptions often go unnoticed in statistical practice, they also sometimes rear their head. For instance, questions such as "In what sense are models selected by AIC considered good?" or "Why do we need to think about 'possible outcomes' in causal inference?" are ontological questions *par excellence*. In each section of this book, we will try to reveal the ontological assumptions that underpin a given statistical method, and consider the implications that the method has on our ontological perspective of the world.

Statistics thus mathematically models the world's structure and expresses it in probabilistic statements. But mathematics and the world are two different things. In order to take such mathematical models as models of empirical phenomena, we must interpret these probabilistic statements in a concrete way. For example, what does it mean to say that the probability of a coin's landing heads is 0.5? How should we interpret the notorious p-value? And what kind of state of affair is represented by the statement that a variable X causes another variable Y? *Semantics*, which is the second warp thread of this book, elucidates the meaning of statements and conceptions that we encounter in statistics.

Statistics is distinguished from pure mathematics in that its primary goal is not the investigation of mathematical structure *per se*, but rather the application of its conclusions to the actual world and concrete problems. For this purpose, it is essential to have a firm grasp of what statistical concepts and conclusions stand for, i.e., their semantics. However, just as statistics itself is not a monolith, so the meaning and interpretation of its concepts are not determined uniquely either. In this book we will see the ways various statistical concepts are understood in different schools of statistics, along with the implications that these various interpretations have for actual inferential practices and applications.

The third and last warp thread of this book is *epistemology*, which concerns the art of correctly inferring the entities that are presupposed and interpreted from actual data. As we noted earlier, statistics is regarded as the primary method by which an empirical claim is given scientific approval in today's society. There is a tacit social understanding that what is "proven" statistically is likely true and can be accepted as a piece of scientific knowledge. What underlies this understanding is our idea that the conclusion of an appropriate statistical method is not a lucky guess or wishful thinking; it is justified in a certain way. But what does it mean for a conclusion to be justified? There has been a long debate over the concept of justification in philosophical epistemology. Similarly, in statistics, justification is understood in different ways depending on the context—what is to be regarded as "(statistically) certain" or counts as statistically confirmed "knowledge" is not the same among, say, Bayesian statistics, classical statistics, and the machine learning literature, and the criteria are not always explicit even within each tradition. This discrepancy stems from their respective philosophical attitudes as to how and why *a priori* mathematical proofs and calculations are able to help us in solving empirical problems like prediction and estimation. This philosophical discordance has led to longstanding conflicts among statistical paradigms, as exemplified by the notorious battle between Bayesians and frequentists in the 20th century. It is not my intention to fuel this smoldering debate in this book; rather, what I want to emphasize is that this kind of discrepancy between paradigms is rooted in the different ways that they understand the concept of justification. Keeping this in mind is important, not in order to decide on a winner, but in order to fully appreciate their respective frameworks and to reflect on why we are able to acquire empirical knowledge through statistical reasoning in the first place. As will be argued in this book, the underlying epistemology of Bayesian statistics and that of classical testing theory are akin to internalism and externalism in contemporary epistemology, respectively. This parallelism, if it holds, is quite intriguing, given the historical circumstance that statistics and philosophical epistemology developed independently without much interaction, despite having similar aims.

With ontology, semantics, and epistemology as our philosophical warp threads, each chapter of this book will focus on a specific statistical method and analyze its philosophical implications; this will constitute the weft of this book.

Chapter 1 is a preliminary introduction to statistics without tears for those who have no background knowledge of the subject. It reviews the basic distinction between descriptive and inferential statistics and explains the minimal mathematical framework necessary for understanding the remaining chapters, including the notions of sample statistics, probability models, and families of distributions. Furthermore, the chapter introduces the central philosophical ideas that run through this book, namely that this mathematical framework represents an ontology for inductive reasoning, and that each of the major statistical methods provides an epistemological apparatus for inferring the entities thus postulated.

With this basic framework in place, Chapter 2 takes up Bayesian statistics. After a brief review of the standard semantics of Bayesian statistics, namely the subjective interpretation of probability, the chapter introduces Bayes' theorem and some examples of inductive inference based on it. The received view takes Bayesian inference as a process of updating—through probabilistic calculations and in accordance with evidence—an epistemic agent's degree of belief in hypotheses. This idea accords well with internalist epistemology, according to which one's beliefs are to be justified by and only by other beliefs, via appropriate inferential procedures. Based on this observation, it will be pointed out that well-known issues of Bayesian statistics, like the justification of prior probabilities and likelihood, have exact analogues in foundationalist epistemology, and that if such problems are to be avoided, inductive inference cannot be confined to internal calculations of posterior probabilities but must be opened up to holistic, extra-model considerations, through model-checking and the evaluation of predictions.

Chapter 3 turns to so-called classical statistics, and in particular the theory of statistical hypothesis testing. We briefly review the frequentist interpretation of probability, which is the standard semantics of classical statistics, and then we introduce the basics of testing theory, including its key concepts like significance levels and *p*-values, using a simple example. Statistical tests tell us whether or not we should reject a given hypothesis, together with a certain error probability. Contrary to a common misconception, however, they by no means tell us about the truth value or even probability of a hypothesis. How, then, can such test results justify scientific hypotheses? We will seek a clue in externalist epistemology: by appealing to a view known as reliabilism and Nozick's tracking theory, I argue that good tests are reliable epistemic processes, and their conclusions are therefore justified in the externalist sense. The point of this analogy is not simply to draw a connection between statistics and philosophy, but rather to shed light on the well-known issues of testing theory. In particular, through this lens we will see that the misuse of *p*-values and the replication crisis, which have been a topic of contention in recent years, can be understood as a problem concerning the reliability of the testing process, and that the related criticism of classical statistics in general stems from a suspicion about its externalist epistemological character.

While the aforementioned chapters deal with classical themes in statistics, the fourth and fifth chapters will focus on more recent topics. The main theme of Chapter 4 is prediction, with an emphasis on the recently developed techniques of model selection and deep learning. Model selection theory provides criteria for choosing the best among multiple models for the purpose of prediction. One of its representative criteria, the Akaike Information Criterion (AIC), shows us that a model that is too complex, even if it allows for a more detailed and accurate description of the world, may fare worse in terms of its predictive ability than a simpler or more coarse-grained model. This result prompts us to reconsider the role of models in scientific inferences, suggesting the pragmatist idea that modeling practices should reflect and depend on the modeler's practical purposes (such as the desired accuracy of predictions) as well as limitations (the size of available data). On the other hand, deep learning techniques allow us to build highly complex models, which are able to solve predictive tasks with big data and massive computational power. The astonishing success of this approach in the past decade has revolutionized scientific practice and our everyday life in many aspects. Despite its success, however, deep learning models differ from traditional statistical models in that much of their theoretical foundations and limitations remain unknown—in this respect they are more like accumulations of engineering recipes developed through trial and error. But in the absence of theoretical proofs, how can we trust the outcomes or justify the conclusions of deep learning models? We will seek a clue to this question in virtue epistemology, and argue that the reliability of a deep learning model can be evaluated in terms of its model-specific epistemological capability, or epistemic virtue. This perspective opens up the possibility of employing philosophical discussions about understanding the epistemic abilities of other people and species for thinking about what "understanding a deep learning model" amounts to.

8 Introduction

Chapter 5 changes gears and deals with causal inference. Every student of statistics knows that causality is not probability—but how are they different? In the language of the present book, they correspond to distinct kinds of entities; in other words, probabilistic inference and causal inference are rooted in different ontologies. While predictions are inferences about this actual world, causal inferences are inferences about possible worlds that would or could have been. With this contrast in mind, the chapter introduces two approaches to causal inference: counterfactual models and structural causal models. The former encodes situations in possible worlds using special variables called potential outcomes, and estimates a causal effect as the difference between the actual and possible worlds. The latter represents a causal relationship as a directed graph over variables and studies how the topological relationships among the graph's nodes determine probability distributions and vice versa. Crucial in both approaches is some assumption or other concerning the relationship between, on the one hand, the data observed in the actual world and, on the other, the possible worlds or causal structures which, by their very nature, can never be observed. The well-known "strongly ignorable treatment assignment" assumption and the "causal Markov condition" are examples of bridges between these distinct ontological levels, without which causal relationships cannot be identified from data. In causal inference, therefore, it is essential to keep in mind the ontological level to which the estimand (the quantity to be estimated) belongs, and what assumptions are at work in the estimation process.

On the basis of these considerations, the sixth and final chapter takes stock of the ontological, semantic, and epistemological aspects of statistics, with a view toward the fruitful and mutually inspiring relationship between statistics and philosophy.

Figure 0.1 depicts the logical dependencies among the chapters. Since philosophical issues tend to relate to one another, the parts of this book are written

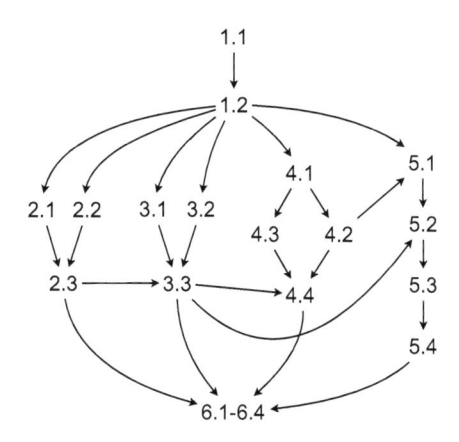

FIGURE 0.1 Flowchart of the book

in such a way that they reflect as many of these organic connections as possible. Readers who are interested in only certain portions of the book will find the diagram useful for identifying relevant contexts and subsequent material. At the end of each chapter I have included a short book guide for the interested reader. I stress, however, that the selection is by no means exhaustive or even standard: rather, it is a biased sample taken from a severely limited pool. There are many good textbooks on both statistics and philosophy, so the reader is encouraged to consult works that suit their own needs and tastes.

1
THE PARADIGM OF MODERN STATISTICS

OK, so let's get down to business. Statistics is, very roughly speaking, the art of summarizing data and using this information to make inferences. This chapter briefly reviews the ABCs of the mathematical framework that underpins these activities. Modern statistics is divided into two parts, *descriptive statistics* and *inferential statistics*, which we will review in turn, laying out their respective philosophical backgrounds. Although the mathematics is kept to the bare minimum, this chapter contains the highest dose of mathematical symbols in the entire book. But there's no need to be afraid: they're not that complicated at all, and understanding the mathematical details, though useful, is not an absolute requisite for following the subsequent philosophical discussions. The most important thing for our purpose is to grasp the ideas behind the mathematical apparatus, so an impatient reader may just skim or skip the formulae on their first reading and return to the details later if necessary.

1.1 Descriptive Statistics

As its name suggests, the historical origin of statistics is closely related to the formation of modern *states*. The development of modern centralized nations in western Europe during the 18th and 19th centuries was accompanied by an "avalanche of printed numbers" (Hacking 1990). In the name of taxation, military service, city planning, and welfare programs, information of all kinds from all over a country was collected by the rapidly emerging bureaucratic system and reported to the central government in the form of printed figures. The flood of numbers confronted policymakers with an urgent need to summarize and extract necessary information from "big data" for decision making. It is still a common practice today to summarize observed data in terms of its mean or variance, or to visualize the data using a plot or histogram. The whole set of such techniques we use to summarize data and make them intelligible is called *descriptive statistics*. The various indices used to summarize data are called *sample statistics* or simply *statistics*, representative examples of which include sample means, sample variances, and standard deviations.

1.1.1 Sample Statistics

Univariate statistics

Imagine there are n students in a classroom, and we represent their height with a variable X. Specific values obtained by measuring students' height are denoted by $x_1, x_2, \ldots, x_n$, where x_i is the height of the ith student, so that $x_i = 155$ if she is 155 cm tall. If, on the other hand, we use another variable Y to denote the age, $y_i = 23$, say, means that the ith student is 23 years old. In general, variables (denoted by capital letters) represent characteristics to be observed, while their values (small letters) represent the results of the observation. A set of observed data is called a *sample*.

The *sample mean* of variable X is the total sum of the observed values of X divided by the sample size n:

$$\bar{X} = \frac{x_1 + x_2 + \cdots + x_n}{n} = \frac{1}{n}\sum_{i}^{n} x_i.$$

The sample mean summarizes data by giving their "center of mass." Another representative index is the *sample variance*, defined as follows.[1]

$$\mathrm{var}(X) = \frac{1}{n}\sum_{i}^{n}(x_i - \bar{X})^2.$$

In order to calculate the sample variance, we subtract the mean from each data point, square the result, and then take their mean (we take the square so that we count positive and negative deviations from the mean equally). Each summand measures the distance of the corresponding data point from the mean; hence, their sum is small if the data are concentrated around the mean, and large if they are scattered widely. The sample variance thus represents the extent of the overall dispersion of the data.

The sample variance in a sense "exaggerates" the deviations because they get squared in the calculation. If one wants to know the dispersion in the original units, one can take the square root of the variance to get the *standard deviation*:

$$\mathrm{sd}(X) = \sqrt{\mathrm{var}(X)} = \sqrt{\frac{1}{n}\sum_{i}^{n}(x_i - \bar{X})^2}.$$

Multivariate Statistics

The sample statistics mentioned in the previous subsection focus on just one aspect or variable of the data. When there is more than one variable, we are sometimes interested in the relationship between them. We may be interested, for example, in whether students' height and age covary, so that older students tend to be taller. The degree to which one variable X varies along with another Y is measured by their *sample covariance*:

$$\mathrm{cov}(X,Y) = \frac{1}{n}\sum_{i}^{n}\left(x_i - \bar{X}\right)\left(y_i - \bar{Y}\right).$$

The idea is similar to variance, but instead of squaring the deviations of X from its mean, for each data point we multiply the deviation of X with that of Y and then take their mean. Since each summand in this case is a product of deviations, one in X and the other in Y, it becomes positive when the variables deviate in the same direction—so that x and y are both above or below their means—and negative when the deviations are in the opposite direction—i.e., when one is above while the other is below their corresponding means. Summing these up, the sample covariance becomes positive if X and Y tend to covary, and negative if they tend to vary in opposite ways.

The covariance divided by the standard deviation of each variable is called the *correlation coefficient*:

$$\mathrm{corr}(X,Y) = \frac{\mathrm{cov}(X,Y)}{\mathrm{sd}(X)\mathrm{sd}(Y)}.$$

The correlation coefficient is always within the range $-1 \leq \mathrm{corr}(X,Y) \leq 1$ and is therefore useful when we want to compare the relative strength of the relationship between a pair of variables with that of another pair. When the correlation coefficient of two variables is larger (or smaller) than zero, they are said to be positively (or negatively) correlated.

Covariance and correlation are symmetric measures of the association between two variables. But sometimes our interest is directional, and in that case it would be useful to relate one variable to another along this direction. We may be interested, for example, in how students' height changes on average when they become a year older. This is given by

$$b_{x,y} = \frac{\mathrm{cov}(X,Y)}{\mathrm{var}(Y)},$$

which is called the *regression coefficient* of X on Y and represents the change in X per unit change in Y. The answer to the aforementioned question therefore becomes: according to the data, the height X increases on average by $b_{x,y}$ for every increase in age Y by a year. The regression coefficient gives the slope of the *regression line*. That is, it is the slope of the line that best fits the data, or, more precisely, the line that minimizes the sum of the squared deviations from each data point.

Discrete Variables

The features we observe need not be expressed in terms of continuous variables, as in the case of height. We may, for example, represent the outcome of a coin flip with the variable X, and let 1 denote heads and 0 tails. Variables like these that do not take continuous values are called *discrete variables*. In this example, the outcome of n coin flips can be represented as a sequence $(x_1, x_2, \ldots, x_n)$, where each x_i is either 0 or 1. The mean $\bar{X}$ is the proportion of heads among the n trials. The sample variance can be calculated in the same way, and it is greatest when one gets an equal number of heads and tails, and zero if only one side of the coin comes up. Thus, the sample variance measures the dispersion of an outcome, just as in the case of continuous variables.

1.1.2 Descriptive Statistics as "Economy of Thought"

Let us pause here and think about what all the statistical quantities we have just defined tell us. As noted at the outset, the primary role of sample statistics is to present large data in an easy-to-understand way, and thereby to uncover structures or relationships that are often invisible when the data are presented as a mere sequence of numbers. This is illustrated by Figure 1.1, the first regression plot made by Francis Galton, the progenitor of regression analysis, in order to study the relationship between the mean parental height (the average of the mother's and father's heights) and the average height of their children among 205 families in 19th-century England. The positive slope of the regression line tells us that the children of taller-than-average parents tend to be tall on average. At the same time, the fact that the slope is less than one indicates that parental height is not perfectly inherited to offspring; in other words the children's height is on average not as "extreme" as their parents' height. Galton named this phenomenon *regression toward the mean* and worried that, just like height, extraordinary talents and qualities of humankind may sink into mediocrity if left to the course of nature. Setting aside the question of whether his worry is justified, Galton's regression analysis vividly illustrates the relationship between parents' and children's heights, or to put it a bit dramatically, it succeeds in uncovering a hidden regularity or law buried in the raw data.

Galton's work well-embodies one of the *Zeitgeister* of his day, namely, the positivist vision of science according to which the objective of scientific investigation is nothing but to summarize and organize the observed data into a coherent order. The core thesis of positivism is that scientific discourse must

14 The Paradigm of Modern Statistics

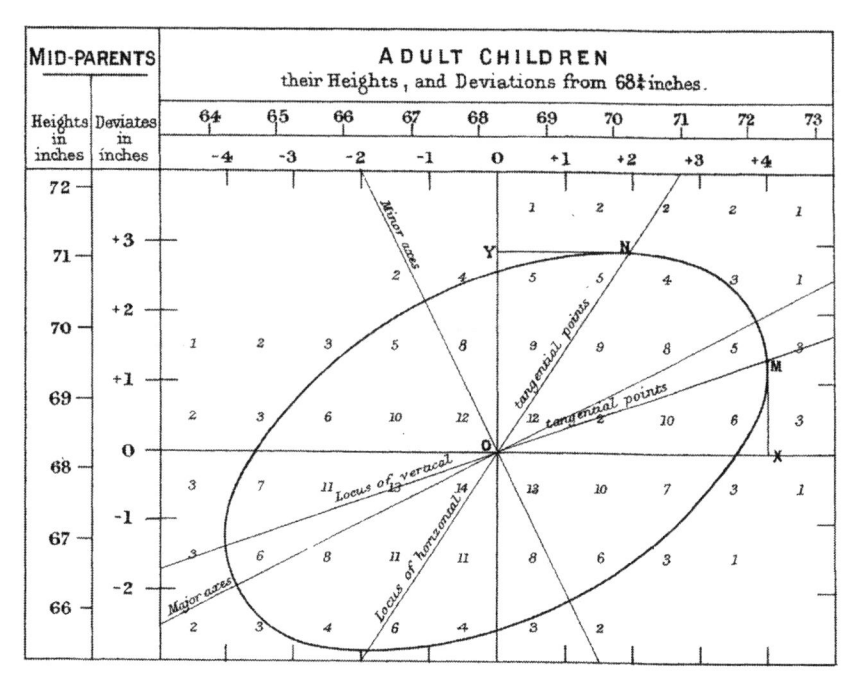

MID-PARENTS		ADULT CHILDREN their Heights, and Deviations from 68¼ inches.								
Heights in inches	Deviates in inches	64 −4	65 −3	66 −2	67 −1	68 0	69 +1	70 +2	71 +3	72 +4	73
72							1	2	2	2	1
71	+3					2	5	5	4	3	1
70	+2		1	2	3	5	8	9	9	8	5 3
69	+1		2	3	6	10	12	12	10	6	3
68	0		3	7	11	14	13	10	7	3	1
67	−1	3	6	8	11	11	8	6	3	1	
66	−2	2	3	4	6	4	3	2			

FIGURE 1.1 Galton's (1886) regression analysis. The vertical axis is the mean parental height and the horizontal axis is the mean children's height, both in inches, while the numbers in the figure represent the counts of the corresponding families. The regression lines are those that are labeled "Locus of vertical/horizontal points."

be based on actual experience and observation. This slogan may sound quite reasonable, or even a truism: surely, we expect the sciences be based on facts alone and refrain from employing concepts of unobservable, supernatural entities such as "God" or "souls"? The primary concern of positivism, however, was rather those concepts *within science* that appear "scientific" at first sight but are nevertheless unobservable. A notable example is the concept of the atom, which was incorporated by Boltzmann in his theory of statistical mechanics in order to explain properties of gases such as temperature and pressure in terms of the motion of microscopic particles and their interactions. Ernst Mach, a physicist and the fiery leader of the positivist movement, attacked Boltzmann's atomism, claiming that such concepts as "atoms" or "forces" are no better than "God" or "souls," in that they are utterly impossible to confirm observationally, at least according to the technological standards of the time. Alleged "explanations" that invoke such unobservable postulates do not contribute to our attaining a solid understanding of nature, and thus should be rejected. Instead of fancying unobservable entities, genuine scientists should devote their efforts to observing data and organizing them into a small number of neat and concise laws so that we can attain a clear grasp of the phenomena—Mach promoted such an "economy of thought" as the sole objective of science.

Mach's vision was taken over by Galton's successor, Karl Pearson, who laid down the mathematical foundation of descriptive statistics. Pearson redirected Mach's positivist attack on unobservables to the concept of causality. Although the idea of a causal relationship, where one event A brings about another event B, is ubiquitous in our everyday conversations as well as scientific discourse, a closer inspection reveals that its empirical purport is hardly obvious. Imagine that a moving billiard ball hits another one at rest, making the latter move. All that one observes here, however, is just the movement of the first ball and a collision, followed by the movement of the second ball. When one says upon making this observation that "the former brings about the latter," one doesn't actually witness the very phenomenon of "bringing about." As Hume had already pointed out in the 18th century, all we observe in what we call a causal relationship is just a *constant conjunction* in which the supposed cause is followed by the supposed effect; we do not observe any "force" or the like between the two. Nor do we find anything like this in numerical data: all that descriptive statistics tells is that one variable X is correlated with another Y, or that the slope of the regression line between them is steep; we never observe in the data the very causal relationship where the former brings about the latter. All this suggests, Pearson argues, that the concept of causality, along with "God" and "atoms," is unobservable and has no place in positivist science, where only those concepts that have a solid empirical basis are allowed. Note that this claim differs from the oft-made remark that causation cannot be known or inferred from correlation. Pearson's criticism is much stronger, in that he claims we should not worry about causality to begin with because science has no business with it. What we used to call "causality" should be replaced by constant conjunction and redefined in terms of the more sophisticated concept of the correlation coefficient, which can be calculated from data in an objective and precise manner. In Pearson's eyes, descriptive statistics provides just enough means to achieve the positivist end of economy of thought, and the concept of causality, being a relic of obsolete metaphysics, must be expelled from this rigid framework.

Ontologically speaking, positivism is an extreme data monism: it claims that the only things that genuinely "exist" in science are data that are measured in an objective manner and concepts defined in terms of these data, while everything else is mere human artifact. Any concept, to be admissible in science, must be reducible to data in an explicit fashion or else banned from scientific contexts, however well it may appear to serve in an explanation. This itself is an old idea that can be traced back to 17th- and 18th-century British empiricism, which limited the source of secure knowledge to perceptual experience alone. But erecting a scientific methodology upon this metaphysical thesis calls for a more rigorous and objective reformulation of its ontological framework. Suppose, for instance, that a follower of Hume claimed that all that exist are constant conjunctions among events, where "constant conjunction" means that those events tend to co-occur. This, however, is at best ambiguous and does not tell us what it means exactly for two events to co-occur, nor does it give us any clue as to how many co-occurrences are sufficient for us to conclude that there is a constant conjunction. Correlation coefficients answer these questions by providing an objective criterion: two variables are related or constantly conjunct when their correlation coefficient is close to one. Of course, there is still some ambiguity as to how close to one it must be, but at least it offers a finer-grained expression that enables one to, say, compare the strength of several conjunctive relationships by looking at their respective correlation coefficients. Descriptive statistics thus furnishes the positivist agenda with a substantive methodology for representing, organizing, and exploring observed raw data, allowing us to extract meaningful relationships and laws. Pearson published his scientific methodology in *The Grammar of Science* (Pearson 1892), a title that boldly declares his manifesto that descriptive statistics, which articulates Mach's data-monistic ontology in a precise language, is *the* canonical approach to positivist science.

1.1.3 Empiricism, Positivism, and the Problem of Induction

So far we have briefly reviewed the basics of descriptive statistics and its methodological implications for the positivist vision of science as a pursuit of economy of thought. One may ask, however, whether this positivist vision accurately represents actual scientific practice and its goals. Driven by the epistemic tenet that knowledge must be built on a secure ground, the positivist philosophy trims down the foundation of science to just those phenomena that are observable and measurable, rejecting all other concepts that are irreducible to experience as metaphysical nonsense. The certainty we derive from this ascetic attitude, however, comes with a high price. The highest is the impossibility of inductive reasoning, already pointed out by Hume. Induction is a type of reasoning that infers an unobserved or unknown state of affairs from given experience, observation, or data. Characterized as such, it encompasses the majority of our inferential practices, from the mundane guesswork about whether one can find an open table in the lunchtime cafeteria to the scientific assessment of the efficacy of a new drug based on clinical trials. In carrying out these inferences, we implicitly assume that the unobserved phenomena to be inferred should be similar to the observed ones that serve as the premise of our inference. Hume called this assumption—that nature operates in the same way across past and future—the *uniformity of nature*. It should be noted that this assumption of uniformity can never be justified by past experiences, because our experiences are just a historical record and by themselves contain no information about the yet-to-be-experienced future. This means that inductive inferences inevitably involve an assumption that cannot be observed within or confirmed by our experience. Hence, if we were to strictly follow the positivist standard and kick out from scientific investigation all such postulates that lack an empirical justification, we at the same time lose all grounds for inductive reasoning.

The same limitation applies to descriptive statistics. In fact, the whole framework of descriptive statistics does not authorize or allow us to make any prediction beyond the data, because prediction lies outside the duties of descriptive statistics, which are to summarize existing data. Galton's regression line in Figure 1.1 by itself says nothing about the height relationships of other families not included in his 205 samples. It is true that we can hardly resist the temptation to guess that those unobserved families, if observed, will also be distributed near the line. But this irresistible feeling, to use Hume's phrase, is just our "mental habit" and has no theoretical or empirical justification. According to the "grammar" of descriptive statistics, such a prediction is no different from, say, a superstitious faith in the "mystical power" of a charm that has survived numerous disasters, and has no place in positivist scientific discourse.

This may be a bold attitude, but as scientific methodology, it is utterly unsatisfactory. Granted, organizing various phenomena into a well-ordered system and uncovering past tendencies are certainly important tasks in science. But we expect a lot more: in particular, we expect science to provide predictions or explanations of unobserved or unobservable phenomena. The pure positivist framework of descriptive statistics falls short in this respect. To capture this predictive and explanatory aspect of scientific practice calls for a more powerful statistical machinery, to which we turn now.

1.2 Inferential Statistics

While descriptive statistics specializes in summarizing observed data, inferential statistics is the art of inferring and estimating unobserved phenomena. As noted earlier, such inductive inferences cannot be justified from the data alone; they must presuppose what Hume called the uniformity of nature behind the data. Inferential statistics formulates this assumed uniformity in terms of a *probability model*,[2] which enables a rigorous and quantitative treatment of inductive reasoning. Figure 1.2 illustrates the overall strategy of inferential statistics. In this framework, data are reconstrued as samples taken from the underlying probability model. Being a random process, each sampling should give a different dataset, while the probability model itself is assumed to stay invariant over the inferential procedure. But since the probability model is by definition unobservable, it must be estimated from the given data, and this estimated probability model serves as the basis for predicting future or unobserved data (illustrated by the dashed arrow in the figure). Inferential statistics thus deals with the problem of induction by introducing the probability model as a new "entity"

18 The Paradigm of Modern Statistics

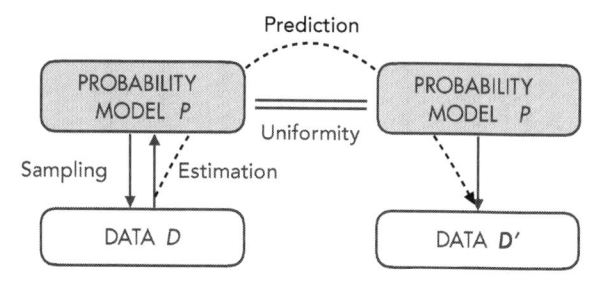

FIGURE 1.2 Dualism of data and probability models. In inferential statistics, data are interpreted as partial samples from a probability model, which is not directly observed but only inferred inductively from the data. The assumption that this probability model remains uniform underlies predictions from the observed to the unobserved. Whereas the concepts of descriptive statistics, i.e., sample statistics, describe the data-world below, those of probability theory (see Section 1.2.1) describe the world above.

behind the data.[3] In other words, it attacks Hume's problem with the dualist ontology of data and models, a richer ontology than that of positivism, which restricts the realm of scientific entities only to data.

But how is this conceptual scheme put into actual practice? Two things are necessary in order for this ontological framework to function as a method of scientific inference. First, we need a mathematical machinery to precisely describe the newly introduced entity, the probability model. Second, we need a definite epistemological method to estimate the model thus assumed from the data. We will focus on the mathematical properties of probability models in the rest of this chapter, and save the latter question for the following chapters.

1.2.1 Probability Models

Probability models are described in the language of *probability theory*. Note that so far the word "probability" has not appeared in our discussion of descriptive statistics in the previous section. This is because the concept of probability familiar in our everyday lives belongs not to data, but to the world behind it from which we (supposedly) take the data. This world-as-source is called a *population* or *sample space*. Roughly speaking, it is the collection of all possible outcomes that can happen in a given trial, observation, or experiment. For instance, the sample space for a trial of rolling a die once consists of $\Omega = \{1, 2, 3, 4, 5, 6\}$, and if we roll it twice, it will be the product $\Omega \times \Omega$. An election forecast, on the other hand, would take all possible voting behaviors of all the voters as its sample space. What we call *events* are subsets of a sample space.

The event of getting an even number by rolling a die is {2, 4, 6}, which is a subset of Ω = {1, 2, 3, 4, 5, 6}; likewise, the event of getting the same number in two rolls is {(1, 1), (2, 2), . . ., (6, 6)} ⊂ Ω × Ω, which again is a subset of the corresponding sample space. In what follows, we denote the sample space by Ω and its subsets (i.e., events) by roman capital letters A, B, and so on. As we will see shortly, a probability measures the size of the "area" these events occupy within the sample space. But due to some mathematically complicated reasons (the details do not concern us here), we cannot count arbitrary subsets as "events," because some are not measurable. There are conditions or rules that a subset must satisfy in order for it to count as a *bona fide* event to which we can assign a specific probability value. The rules are given by the following three axioms:

R1 The empty set ∅ is an event.
R2 If a subset $A \in \Omega$ is an event, so is its compliment $A^c = \Omega/A$.
R3 If subsets $A_1, A_2, \ldots$ are events, so is their union $\cup_i A_i$.[4]

Nothing complicated—these rules require merely that if something is an event, then its negation must also be an event, and if there are multiple events, then their combination (union) must also be regarded as an event. A set of events that satisfies these conditions is called a σ-algebra, but we do not have to worry about it in this book. In the previous example of rolling a die, the power set (i.e., the set of all subsets) of the sample space Ω gives a σ-algebra satisfying the aforementioned three conditions, and in this case any subset in the sample space counts as an event.[5]

The probability of an event, as mentioned earlier, is its "size" in the sample space. The "size" is measured by a *probability function* P that satisfies the following three axioms.[6]

Probability Axioms

A1 $0 \leq P(A) \leq 1$ for any event A.
A2 $P(\Omega) = 1$.
A3 If events $A_1, A_2, \ldots$ are mutually exclusive (i.e., they do not overlap), then
$$P(A_1 \cup A_2 \cup \ldots) = P(A_1) + P(A_2) + \ldots.$$

Axiom 1 states that the probability of any event (a subset of the sample space) falls within the range between zero and one. By Axiom 2, the probability of the entire sample space is one. Axiom 3 stipulates that the probability of the union of non-overlapping events/subsets is equal to the sum of their probabilities (this axiom justifies our analogy of probabilities and sizes).

A probability model consists of the aforementioned three elements: a sample space, a σ-algebra defined on it, and a probability function. This triplet is all there is to probability. From this we can derive all the theorems of probability theory, including the following elementary facts, which the reader is encouraged to verify:

T1 $P(A^c) = 1 - P(A)$, where A^c is the compliment of A.
T2 For any events A, B (which need not be mutually exclusive), $P(A \cup B) = P(A) + P(B) - P(A \cap B)$.

That is, the probability of "A or B" is equal to the sum of the probability of A and that of B minus that of their intersection (i.e., the probability of "A and B"). For simplicity, we hereafter write $P(A, B)$ instead of $P(A \cap B)$.

Conditional Probability and Independence

Although we said that the aforementioned axioms define all there is to probability, a few additional definitions will prove useful in later discussions. The *conditional probability* of A given B is defined by:

$$P(A \mid B) = \frac{P(A,B)}{P(B)}.$$

We may think of a conditional probability as the probability of some event (A) given that another event (B) has occurred. In general, a probability of an event changes by conditioning. But when it doesn't, so that $P(A \mid B) = P(A)$, we say that A and B are *independent*. Independence means irrelevance: information about B gives us no new information about A when they are independent. When A and B are not independent, they are said to be *dependent*. The independence relation satisfies the following properties, which the reader should verify using the definition just provided:

- Symmetry, i.e., $P(A \mid B) = P(A)$ iff (if and only if) $P(B \mid A) = P(B)$.
- If A and B are independent, then $P(A, B) = P(A) P(B)$; that is, the probability that they both hold is the product of each probability.

Marginalization and the Law of Total Probability

Suppose events $B_1, B_2, \ldots, B_n$ partition the sample space; that is, they are mutually exclusive (i.e., $B_i \cap B_j = \emptyset$ for $i \neq j$) and cover the entire sample space when combined $\left(\cup_i^n B_i = \Omega\right)$. Then, for any event A, we have

$$P(A) = \sum_i^n P(A, B_i).$$

That is, A's "area" can be reconstructed by patching together all the places where A and B_i overlap. This is called *marginalization*. The terms on the right-hand side can be rewritten using conditional probabilities to yield

$$P(A) = \sum_i^n P(A|B_i)P(B_i),$$

which is called the law of total probability.

Bayes' theorem

By rearranging the right-hand side of the definition of conditional probability, we obtain

$$P(A|B) = \frac{P(B|A)P(A)}{P(B)}$$

for any events A, B. This equation, called Bayes' theorem, plays an essential role in Bayesian statistics, which we will discuss in the next chapter.

1.2.2 Random Variables and Probability Distributions

As stated earlier, events are subsets of a sample space. These events, as they stand, so far have no "name" and have been simply denoted by the usual subset notation, so that the probability of the event of getting an even number by rolling a die, say, is denoted as $P(\{2, 4, 6\})$. Enumerating all the elements in this way may not cause any inconvenience in this particular example, as it has only three possibilities; but it may be very cumbersome when dealing with a much larger event, say, when we want to refer to all eligible voters in the sample space of all citizens. To resolve this issue, statisticians use *random variables* to identify an event of interest within the whole sample space. Random variables are (real-valued) functions defined on a sample space to indicate properties of objects. For example, let the random variable Y be a function that gives the age of a person. Then Y (Homer Jay Simpson) $= 40$ means that Mr. Simpson in the sample space is 40 years old. Now let's assume that in a certain country voting rights are given to citizens who are 18 years of age or older. Then, using the aforementioned function, the subset "eligible voters" can be expressed as the inverse image $\{\omega \in \Omega : Y(\omega) \geq 18\}$, that is, the set that consists of all elements ω in the sample space Ω for which Y gives a value equal to or greater than 18. But since this notation is still lengthy, we simply write $Y \geq 18$ to denote the subset. Likewise, if X represents height, then $X = 165$ stands for $\{\omega \in \Omega : X(\omega) = 165\}$ and refers to the set of people who are 165 cm tall.[7] Now, imagine a trial where we randomly pick a subject from the whole population of citizens. With this setup and our definition of events as subsets of the sample space, $Y \geq 18$ corresponds to the event that the selected person is no less than 18 years old, while $X = 165$ corresponds to the event that he or she is 165 cm tall.

Now, since the event identified by a value of a random variable is a subset of the sample space, we can assign to it a probability. The probability that the selected person is no less than 18 years old is $P(Y \geq 18)$, while the probability that she is 165 cm tall is $P(X = 165)$. In general, the probability that a random variable X has value x is given by $P(X = x)$. With this notation and setup, $P(X = 165) = 0.03$ means that the probability of selecting a person who is 165 cm tall is 3%. Do not get confused by the double equal signs: the first one, $X = 165$, is just a label telling us to pick out the event that the selected person is 165 cm tall, while the second equal sign is the one that actually functions as an equality connecting both sides of the equation. But since this expression is rather repetitive, we sometimes omit the first equal sign and simply write $P(x)$ to denote $P(X = x)$ when the relevant random variable is clear from the context. In this shorthand, $P(x) = 0.01$ means that the probability of X having value x is 1%. We can also combine two or more variables to narrow down an event. For example, $P(x, y)$ stands for $P(X = x \cap Y = y)$, and according to the prior interpretation, it denotes the probability of selecting a person who is x cm tall *and* y years old.

What is the point of introducing random variables? In most statistical analyses, we are interested in attributes or properties of objects, like height or age. Representing these attributes in terms of random variables allows us to express how the probability depends on the value of these properties; in other words, it allows us to reconstrue the probability function as a function that assigns probability values to specific properties rather than events. The function that assigns probability $P(x)$ to value x of random variable X is called a *probability distribution* of X and is denoted by $P(X)$. Note the difference between the uppercase X and lowercase x. $P(X)$ is a *function* that can be represented by a graph or histogram that takes X as its horizontal axis and the probability values as its vertical axis. On the other hand, $P(x)$ or $P(X = x)$ (recall that the former is just a shorthand of the latter) is a particular probability *value* that the function returns given the input x, and it is represented by the height of the function/graph $P(X)$ at x.

When there are two or more random variables X and Y, a function that assigns the probability $P(x, y)$ to each combination of their values x, y is called a *joint probability distribution* of X and Y. This can be illustrated by a three-dimensional plot where $P(x, y)$ is the height at the coordinate (x, y) on the horizontal plane $X \times Y$. The joint distribution $P(X, Y)$ contains all the probabilistic information about the random variables X and Y. To extract the information of just one variable, say Y, we marginalize it by taking the sum[8] over all values of X:

$$P(y) = \sum_x P(y, x).$$

Calculating this probability for each $Y = y$ yields the distribution of Y, called a *marginal probability distribution*.

Probability distributions are essentially the same as the probability functions defined in the previous section; in fact, $P(x)$ is simply the probability value of the event identified by $X = x$. Hence, all the axioms, definitions, and theorems of probability theory carry over to probability distributions. For instance, the conditional probability that someone has height x given that she is y years old is defined as $P(x|y) := P(x, y)/P(y)$. Likewise, when $P(x|y) = P(x)$, or equivalently, $P(x, y) = P(x)P(y)$, the two events $X = x$ and $Y = y$ are independent. This is a relationship that holds between particular values x and y. More generally, when the independence condition $P(x, y) = P(x)P(y)$ holds for *all* values x, y of the random variables X, Y, these variables are said to be independent. Note the difference between the two: the independence of values is a relationship between concrete events, whereas the independence of random variables is a general relationship between attributes. In terms of the previous example, the former only claims that knowing that a person is y years old does not give any extra clue as to whether she measures x cm or not, whereas the latter is the much stronger claim that knowing a person's age gives no information whatsoever about her height—which sounds unlikely in this particular example, but may hold for other attributes, such as height and commuting time.

Note on Continuous Variables and Probability Densities

As we have seen in Section 1.1.1, some attributes take discrete values, while others take continuous values. Properties like height that vary gradually may be better represented by continuous variables that take real values. Continuous random variables, however, require a bit of caution when we consider their probability. Suppose X is a continuous random variable: what is the probability that it takes a specific real value x, i.e., $P(X = x)$? Whatever x is, the answer is always zero. To see why, recall that a probability function is a measure of the size of a subset in a sample space. An element among uncountably many elements, like a point on a real line, does not have any extension or breadth, and so its "size" or probability must be also zero. This may make intuitive sense if one notes that no one has a height exactly equal to 170.000 . . . cm, no matter large a population we take. Thus, any particular value of a continuous variable has zero probability. But even in the continuous case, a certain interval, say between 169 and 170 cm, may have a nonzero size/probability. We can then consider the result of successively narrowing down this interval. The probability of an infinitely small interval around a point is called a *probability density*, and a function that gives the probability density at each point x is called a *probability density function*. The probability of an interval can be obtained by integrating this function over the interval. Letting f be the probability density function of height, the probability that a person's height falls within the range from 169 to 170 cm is given by:

$$P(169 \leq X \leq 170) = \int_{169}^{170} f(x)dx.$$

Hence, strictly speaking, one should use probabilities for discrete random variables and probability densities for continuous ones. Nevertheless, in this book we will abuse terminology and notation, using the word "probability" to denote both cases, and the expression $P(X = x)$ to denote the probability of x when X is discrete and the probability density at x when X is continuous. Readers who are inclined toward rigor should reinterpret terminology and notation according to the context.

Expected Values

A probability distribution, as we saw, is a function of the values of a random variable. Since this pertains to the "uniformity of nature" that cannot be observed in the data, the entirety of this function is never fully revealed to us (recall that a sample space contains not just what happened or is observed but all the possible situations that might occur). Nevertheless, one can consider values that summarize this "true" distribution. Such values that are characteristics of the probability distribution of a given random variable are called its *expectations* or *expected values*.

A representative expected value is the *population mean*, often denoted by the Greek letter μ and defined by:

$$\mu = \sum_x x \cdot P(X = x)$$

The population mean is the sum of the values x of X weighted by their probability $P(X = x)$, and gives the "center of mass" of the distribution. In contrast, its dispersion is given by the *population variance* σ^2:

$$\sigma^2 = \sum_x (x - \mu)^2 \cdot P(X = x)$$

Expressed in English, this is the sum of the squared deviation $(x - \mu)^2$ of each value x of X from its population mean, weighted by the probability $P(X = x)$.

More generally, an expected value of a random variable is defined as a weighted sum or integral of its values (as in the case of the population mean) or a function thereof (as in the case of the population variance, where the function in question is the squared deviation). This operation is called "taking the expectation" and is denoted by $\mathbb{E}$. For instance, the population mean is the result of taking the expectation of X itself, i.e., $\mathbb{E}(X)$, while the population variance is the result of taking the expectation of $(X - \mu)^2$, i.e., $\mathbb{E}\left[(X - \mu)^2\right]$. Other expected values can be defined in the same manner.

At first sight, the population mean and variance may look very much like sample statistics, such as the sample mean and variance reviewed in Section 1.1.1. They are indeed related, yet they are different in important ways. The primary difference lies in the kind of objects they purport to describe. Recall that sample statistics such as the sample mean are summaries of observed data or samples. In contrast, what we are dealing with here is not the finite data at hand but rather their source, defined as a probability distribution on a certain sample space. The population mean and variance describe this probability model—and in this sense they can be thought of as extensions of the concepts of the sample mean and variance, redefined on the entire sample space which includes not only observed but also unobserved, and even unobservable, samples. Since we cannot measure these samples, expected values, by their very nature, cannot be known directly. They are, so to speak, only in the eyes of God who is able to see all there is to know about a probability model.

The IID Condition as the Uniformity of Nature

That sums up our brief review of the probability model as a "source of data." In order to go beyond the given data and make inferences about unobserved phenomena, inferential statistics needs to posit a uniform structure behind the data. The concepts introduced in this section, such as the sample space, probability functions, random variables, probability distributions, and expected values, are mathematical tools we use to describe or characterize this posited structure, namely the probability model.

This posited structure, however, is still an assumption. In order to make use of it in inductive reasoning, inferential statistics must identify this structure. How do we do this? As previously stated, in inferential statistics the data we observe are interpreted as samplings from the probability model. A crucial assumption in this process is that all the samples come from the same probability model. That is, each piece of data must follow the same probability distribution. Another common requirement besides this is that the sampling must be random: one should not, for example, disproportionately pick taller individuals, or alternate between picking tall and short individuals. This amounts to the requirement that random variables of the same type must be independent,[9] so that one cannot predict what will come next based on any particular outcome (of course, distinct random variables like height and age need not be independent). When these conditions are satisfied, a set of random variables is said to be *independent and identically distributed*, or IID for short.

The IID condition is a mathematical specification of what Hume called the uniformity of nature. To say that nature is uniform means that whatever circumstance holds for the observed, the same circumstance will continue to hold for the unobserved. This is what Hume required for the possibility of inductive reasoning, but he left the exact meaning of "sameness of circumstance" unspecified. Inferential statistics fills in this gap and elaborates Hume's uniformity condition into the more rigorous IID condition.[10] In this reformulation, uniformity means that the probability model remains unchanged across observations, and that the sampling is random so that the observation of one sample does not affect the observation of another. Note that this kind of mathematical formulation is possible only after the nature of the probability model and its relationship with the observed data are clearly laid out. In this way, probability theory provides us with a formal language that enables us to specify the ontological prerequisites of inductive reasoning in mathematical terms.

The Law of Large Numbers and the Central Limit Theorem

The assumption of IID as a uniformity condition allows us to make inferences about the probability model behind the data. This is most vividly illustrated by the famous law of large numbers and the central limit theorem, both of which are part of *large sample theory*, the backbone of traditional inferential statistics.

Let us look at the law of large numbers first. We are interested in estimating, on the basis of observed data, the underlying probability distribution or its expected values like the population mean or variance. Suppose, for example, that we are interested in the mean national height. What we are able to know, however, are only the sample statistics obtained from a finite set of data, such as the sample mean. As we have noted, such sample statistics are distinct from the expected values of the probability distribution, both conceptually and in the way they are defined. Nevertheless, it seems to be a very natural idea to take the latter as an *estimator* of the former. Of course, the mean height of just a handful of people will hardly serve as a reliable estimator; but if we have more data and measure the height of, say, millions of people, we would feel very safe in regarding the sample mean of this data to be a good approximation to the true national mean. The law of large numbers backs up this common-sense intuition with a rigorous mathematical proof that the sample mean will approach the real mean of the population as we collect more and more data. Let the random variables $X_1, X_2, \ldots, X_n$ be IID, i.e., they are mutually independent and have the same distribution. A typical example is the height of n people from the same population. Since these variables are IID, they have the same population mean $\mathbb{E}(X_1) = \mathbb{E}(X_2) = \cdots = \mu$. The law of large numbers then states that as the sample size n approaches infinity, the sample mean $\bar{X}_n = \sum_i^n X_i / n$ *converges in probability* to the population mean μ: that is,

$$\lim_{n \to \infty} P\left(\left| \bar{X}_n - \mu \right| \geq \epsilon \right) = 0$$

holds for an arbitrarily small positive margin ϵ. The probability function on the left-hand side expresses the probability that the deviation of the sample mean $\bar{X}_n$ from the population mean μ is no less than ϵ. The equation as a whole thus states that this probability becomes zero as n approaches infinity, i.e., that it will be certain that the sample and population means coincide with arbitrary precision. This mathematical result provides us with the ground for roughly identifying a sample mean obtained from numerous observations with the population mean.

Note that our only assumption was that $X_1, X_2, \ldots, X_n$ are IID; there was no restriction on the form of their distributions. This means that what is crucial in the law of large numbers is merely the presence of uniformity, and not the specific form of this uniformity: even if the underlying distribution is completely unknown, the mere fact that the samples are IID or uniform ensures that, as we have more and more data, the probability distribution of the sample mean will fall within a certain range around $\bar{X}_n$. But that's not all: the distribution of the sample mean tends toward a unique form, the "bell-shaped" normal distribution. This is the result of the famous *central limit theorem*. Let $X_1, X_2, \ldots, X_n$ be IID variables again, this time with the population variance σ^2. Then the theorem states that as n approaches infinity, the distribution $P(\bar{X}_n)$ of the sample mean tends toward the normal distribution with mean μ and variance σ^2/n. We will return to the normal distribution after explaining the concept of distribution families; all we need to know now is that it is a particular form of distribution. Hence, the result here means that even though we do not know anything about the nature of the underlying distribution, we can know that its sample means tends toward a particular distribution as we keep sampling. What is important here is that this result is derived from the IID condition alone.[11] As in the law of large numbers, all that is required for the central limit theorem to hold is that the data are obtained from a uniform IID process; we need not know about the form or nature of the underlying distribution. From this assumption alone, we can conclude that the sample mean will always converge to the same form, namely the normal distribution. This means that we can make inferences about the unknown probability model just by repeated sampling. In this way, the results of large sample theory such as the central limit theorem and law of large numbers provide us with a theoretical justification of the agenda of inferential statistics, which is to make inductive inferences about the true but unobservable distribution on the basis of finite and partial observations.

1.2.3 Statistical Models

What Statistical Models Are

Let us take stock of our discussion so far. In order to build a framework for inductive inference, a kind of inference that goes beyond the given data, we began by defining a probability model, which is the uniformity behind the data. Then we introduced random variables as a means of picking out the events corresponding to properties that we are interested in, and saw that they have certain definite distributions. Although these distributions are unknown and unobservable, they can be characterized in terms of expected values, and we saw that these expected values can be approached, with the aid of large sample theory, through repeated samplings that satisfy the IID condition.

But this doesn't settle all the inductive problems. Large sample theory provides us only with an eschatological promise, so to speak: it guarantees only that *if* we keep collecting data indefinitely, then the true distribution will eventually be revealed. Hence, *if* we could keep tossing the same coin infinitely many times, the observed ratio of heads would converge to the true probability. It is impossible, however, to actually conduct an infinite number of trials, since the coin will wear out and vanish before we complete the experiment. "In the long run we are all dead," as Keynes said, and given that we can collect only finite and in many cases modest-sized data which fall far short of "large samples" in our lifetime (or the time we can spend on a particular inductive problem), a statistical method, if it is to be serviceable, must be able to draw inferences from meager inputs. Such inferences may well be fallible and may not give probability-one certainty as the law of large numbers does. What becomes important, therefore, is a framework that enables us to carry out inductive inferences as accurately as possible and to evaluate the reliability of these inferences, even within the bounds of limited data. The true value of inferential statistics consists in the development and elaboration of such earthy statistical methods.

To achieve this goal, inferential statistics introduces additional assumptions beyond the basic setup of probability models. Our approach to inductive reasoning up until now relied on a probability model and IID random variables, but made no assumptions about the nature or type of the distribution that these variables have. Or, in Humean parlance, so far we only required the presence of the uniformity of nature, without caring about what nature is like. In contrast, the more realistic inference procedure we are now after makes some further assumptions about the form of the distribution, thereby narrowing down the range of possible distributions to be considered. So-called parametric statistics, for instance, deals only with those distributions that can be explicitly expressed in terms of a particular function determined by a finite number of *parameters*.[12]

It therefore requires not only the existence of uniformity in the form of a probability model, but also that the uniformity is of a certain pre-specified kind (that is, it has a specific functional form). A set of candidate distributions circumscribed in this way is called a *statistical model*. It is important not to confuse statistical models with probability models, as the distinction will be essential in understanding inferential statistics. Probability models, we may recall, were introduced in order to describe the reality that supposedly exists behind data, using the language of probability theory such as sample spaces, σ-algebras, probability functions, random variables, and probability distributions. A statistical model, on the other hand, is just a hypothesis about the nature of the probability distribution that we posit, and may well be regarded as fictional. The true probability distribution may be highly complex, hardly specifiable by a simple function with only a few or even a finite number of parameters. Nonetheless, we pretend for our immediate purposes that it has a particular functional form, and we proceed with our inferential business under this hypothesis.

At this point an attentive reader may raise their eyebrows and ask: "But wasn't a probability model or uniformity of nature also a hypothesis, not amenable to empirical confirmation? If that's the case, then probability models and statistical models are both hypotheses after all, so there doesn't seem to be any substantive distinction between them." It is true that, if we had to say whether they are realities or assumptions, then they are both assumptions. They differ, however, as to what kind of thing they are assumed to be. A probability model or uniformity of nature is assumed to be true: as long as we are to make any inductive inferences, we cannot but believe in the existence of some uniformity—that was Hume's point. This is not so, however, with statistical models. In fact, most statisticians do not believe that any statistical model faithfully describes the real world; rather, they expect it to be only an approximation—that is, a model—good enough for the purpose of solving their inductive tasks. In other words, a statistical model is not assumed to be a true representation of the world, but an instrument for approximating it. This instrumental nature of statistical models is aptly captured by the famous aphorism of the statistician George Box: "all models are wrong, some are useful." In contrast, the very existence of a probability model cannot be a fiction; otherwise, we would be trapped in Humean skepticism and lose all grounds for inductive reasoning.[13]

Parametric Statistics and Families of Distributions

There are two ways to build statistical models. So-called nonparametric statistics makes fairly general and weak assumptions such as the continuity or differentiability of the target distribution, without imposing any specific functional form on it. Parametric statistics, on the other hand, goes a step further and specifies the general form of the distribution in terms of a particular function. Such a distributional form is called a *family of distributions*. Once a family is fixed, individual distributions are uniquely determined by specifying its parameters. This means that parametric statistics models the target only within the range of a fixed family of distributions. It therefore has a higher risk of distorting reality, but allows for more precise and powerful inferences if an appropriate distributional family is chosen. The question of which family to use depends on the nature of the inductive problem and random variables we are interested in. Since this

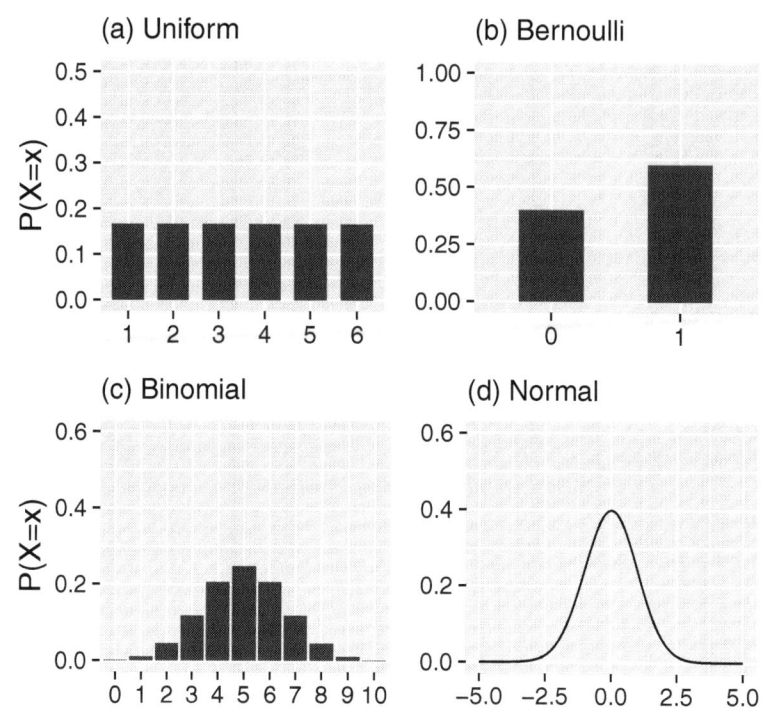

FIGURE 1.3 Examples of distributions. The horizontal and vertical axes represent values of random variable X and their probabilities, respectively. (a) Uniform distribution that assigns the same probability to all faces of a die. (b) Bernoulli distribution with $\theta = 0.6$. (c) Binomial distribution with $\theta = 0.5$, $n = 10$. (d) Normal distribution with $\mu = 0$, $\sigma^2 = 1$.

book will mostly focus on parametric statistics, it should be useful at this point to look at some of the representative distributional families (see also Figure 1.3).

Uniform distribution

A *uniform distribution* assigns the same probability to all values $x_1, x_2, \ldots$ of a random variable X. For example, if X is the result of rolling a fair die, we have a uniform distribution with $P(X = x) = 1/6$ for $x = 1, 2, \ldots, 6$. When X is a continuous variable ranging from α to β, its uniform distribution is

$$P(X = x) = \frac{1}{\beta - \alpha},$$

and is uniquely determined by the parameters α, β. In this book we follow the common practice of denoting the parameters of a distribution by Greek letters, and variables by Roman letters.

Bernoulli distribution

Let the random variable X stand for the outcome of a coin toss, with $X = 0$ representing tails and $X = 1$ heads. If the probability of landing heads is $P(X = 1) = \theta$, the distribution of X is

$$P(X = x) = \theta^x (1-\theta)^{1-x}.$$

Note that $x = 0$ or 1, so the right-hand side becomes $1 - \theta$ if $x = 0$ (tails) and θ if $x = 1$ (heads). In this case, X is said to follow a *Bernoulli distribution*. The mean and variance of a Bernoulli distribution are θ and $\theta(1 - \theta)$, respectively, and the distribution is determined by the single parameter θ.

Binomial distribution

Next, consider an experiment where we toss the same coin 10 times and count the number of heads. Let X be the number of heads we get in these 10 trials. What is its distribution? To answer this question, we examine the probability of each possible outcome in 10 tosses, i.e., the probability of no heads, 1 head, 2 heads, and so on. First, the probability of $X = 0$ (no heads) is

$$(\text{Probability of heads})^0 (\text{Probability of tails})^{10} = \theta^0 (1-\theta)^{10} = (1-\theta)^{10}.$$

Next consider $X = 1$. The probability of getting heads in only the first toss and tails thereafter is

$$(\text{Probability of heads})^1 (\text{Probability of tails})^9 = \theta^1 (1-\theta)^9 = \theta(1-\theta)^9.$$

Now this is the same as the probability of getting heads in just the second, third, ..., or the 10th toss. Thus, by summing up all 10 of these cases, we obtain the probability of getting 1 head out of 10 trials. In general, the probability of getting x heads is

$$P(X = x) = {}_{10}C_x \theta^x (1-\theta)^{10-x},$$

where ${}_{10}C_x = \dfrac{10!}{x!(10-x)!}$ is the number of ways of choosing x out of 10 things. Replacing the 10 in the preceding equation with n gives us the probability of getting x heads out of n coin tosses.

The distribution derived in this way is called a *binomial distribution*. The mean and variance of a binomial distribution are known to be $n\theta$ and $n\theta(1 - \theta)$, respectively. That is, the binomial distribution is completely determined by the probability of each trial θ and the total number of trials n.

Normal distribution

The *normal distribution* is arguably the most famous family of distributions, and we have already encountered it in Section 1.2.2. Unlike Bernoulli and binomial distributions, the normal distribution concerns a continuous variable and is expressed by the equation

$$P(X = x) = \frac{1}{\sqrt{2\pi\sigma^2}} \exp\left\{-\frac{(x-\mu)^2}{2\sigma^2}\right\},$$

with two parameters μ and σ^2, which respectively define the mean and variance of the distribution. Though not obvious at first sight, plotting this equation generates the well-known bell curve, symmetric about the mean.

We have seen that the probability of getting a particular number of heads in multiple coin tosses is given by a binomial distribution. What happens if we continue tossing and increase the number n of tails? It will still be a binomial distribution with large n, but lo and behold, its shape comes closer to that of a normal distribution. This effectively exemplifies the central limit theorem discussed earlier. In effect, a binomial distribution is nothing but the distribution of the sum of repeated Bernoulli trials; hence, as the number of repetitions increases, it asymptotically approaches a normal distribution. As previously stated, the same result holds not just for Bernoulli distributions but also for basically any distribution. From this we can expect that a property determined by the cumulative effects of a number of small factors will follow a normal distribution—typical examples being height and weight, which are presumably affected by numerous genetic as well as environmental factors. Partly for this reason, the normal distribution is very general and plays an important role in statistics.

Multivariate Normal Distribution

Finally, let us take one example from bivariate joint distributions. If two random variables X, Y both follow normal distributions, their joint distribution is called a *multivariate normal distribution* and is expressed by

$$P(X = x, Y = y) = \frac{1}{2\pi\sigma_X\sigma_Y\sqrt{1-\rho^2}} \exp\left\{-\frac{1}{2(1-\rho^2)}\left(\frac{(x-\mu_X)^2}{\sigma_X^2} + \frac{(y-\mu_Y)^2}{\sigma_Y^2} - \frac{2\rho(x-\mu_X)(y-\mu_Y)}{\sigma_X\sigma_Y}\right)\right\}.$$

This may look a bit scary, but there's no need to be afraid. The key is to see that the right-hand side is a function of x and y. This distribution has five parameters, the mean μ_X and variance σ_X^2 of X, the mean μ_Y and variance σ_Y^2 of Y, and the population correlation coefficient ρ of X and Y. These five parameters specify the functional form and determine how the probability value (or, to be precise, probability density) depends on the values x, y.

The means and variances, as we have seen, respectively give the centers of mass of and dispersions of X, Y. The population correlation coefficient ρ, in contrast, represents the strength of association between X and Y within the range from -1 to 1. Although this may look similar to the correlation coefficient described earlier as descriptive statistics, they differ in crucial respects. The point here is similar to the cautionary remark made previously about expected values. The correlation coefficient understood as a sample statistic, as we have seen, measures an association among observed data and is explicitly calculable. In contrast, the population correlation coefficient concerns the underlying probability distribution and is supposed to capture the correlation inherent in the entire population, including unobserved items. As such, the population correlation coefficient, and in general any other parameter, denotes—or, more precisely, models (because a statistical model is a model of the underlying distribution)—a hypothesized property of the population behind the data, but is not something visible or observable. Thus, sample statistics and parameters belong to different "ontological levels," as it were, and for this reason should not be confused.

The families of distributions mentioned in this section are just a tiny sample. Other representative distributional families include the Poisson, exponential, chi-squared, t, and beta distributions, to name just a very few. Information about these and other distributions can be found in any standard textbook or website.

Presupposing a particular family of distributions significantly facilitates statistical inference. As stated earlier, most statistical inferences involve, in one way or another, estimations of the probability distribution behind the data. Hence, if the target distribution can be assumed to belong to a particular distributional family, the whole inferential task is reduced to an estimation of a finite number of parameters. Once a particular variable, say height, is known to have a normal distribution, its statistical properties can be completely determined just by estimating its mean and variance. (If, in contrast, all we can assume is that the distribution is continuous and smooth, we need to estimate infinitely many parameters to attain a complete picture of it.) Parametric statistics thus translates statistical hypotheses into those about the parameters of a distribution, and then solves the inductive problem by inferring these parameters from the data.

Specifying the functional form of a distribution brings with it another important benefit: it enables us to explicitly calculate the *likelihood*, or the probability of obtaining the data under various hypothetical parameters. For instance, given that heights follow a normal distribution, the probability that someone's height is between, say, 145 and 150 cm can be calculated by integrating the aforementioned normal distribution function with a specific mean and variance. In the following chapters, we will see that this concept of likelihood is ubiquitous and plays a central role in Bayesian statistics, hypothesis testing, model selection, and machine learning.

1.2.4 The Worldview of Inferential Statistics and "Probabilistic Kinds"

This concludes our brief review of the theoretical framework of inferential statistics. The framework introduced in this chapter serves as the ontological foundation for inferential statistics. That is, it tells us what kinds of entities must be postulated and how they should be described in order to carry out inductive inferences using statistical methods. But why use probability theory to express such ontological machinery? This is because the mathematical formalism of the theory enables us to assess the ontological assumptions that are necessary for tackling a given inductive task. The minimum condition for inductive inference is the assumption of a probability model or uniformity of nature, which, along with the law of large numbers, guarantees that enumerative induction[14] will eventually converge to the correct answer in the long run. If we are more realistic and want to make an inference based on a finite sample and evaluate its accuracy, the target distribution must be specified in terms of a statistical model. And if we want to make more effective predictions or inferences based on a limited sample size, we need to resort to parametric statistics and specify the type of uniformity we postulate in terms of a particular family of distributions. In this way, stronger ontological assumptions allow for more effective inductive inferences, and this becomes most explicit in the mathematical formulation.

Ever since Hume, philosophers have devoted their attention almost exclusively to the validity of the most fundamental of these ontological conditions, i.e., the question of whether there exists a uniformity of nature. This basic premise, however, is rarely doubted by statisticians. They are rather concerned with the nature of the uniformity, that is, the assumption of a particular distributional family. The appropriate type of regularity or uniformity should differ according to the problem and object under study. At the same time, the same uniformity assumption may be applicable to situations of a quite different nature. There seems to be nothing in common, at least from a physical perspective, between a coin toss and the genital formation of fetuses. Nevertheless, their uniformity can be modeled by the same Bernoulli family of distributions (albeit perhaps with different parameters), so we can use the same kind of distribution to predict which team kicks off first and whether a newborn baby will be a girl or boy. Nay, it would be more appropriate to say that it is through the lens of the Bernoulli distribution that we can regard these physically different things as the same kind of *thing* in the context of inductive inference. On the other hand, while rolling a die seemingly has much more in common with tossing a coin than with human development, it calls for a different distributional family from the coin toss; that is, statistically speaking, they are regarded as different kinds of *things*. Likewise, other families of distributions, such as binomial or normal distributions, represent different types of uniformity. There are varieties of regularity and uniformity in nature: distributional families define these "varieties" in a functional form and classify various inductive problems into specific categories, so that tasks within each category are regarded as the "same kind" of statistical problem, even if they are numerically or physically distinct. In this sense, the choice of a particular family of distributions reflects a modeler's ontological assumption, as it determines what kind of "thing" the task under study is.

Turning our eyes to our everyday cognitive processes, we realize that we rarely, if at all, see the world as it is "in itself," whatever that means. Rather, the world always appears to us as already articulated, segmented into discrete objects. In front of me now are a desk, a laptop, and a bookshelf. Turning to the window, I see the university clock tower, on which a crow is perched. I thus recognize the world as consisting of the desk, the laptop, the bookshelf, the clock tower, the crow, among other things. Each of these things, carved out from the world by my cognitive processes, has its own particular properties, on the basis of which I am able to make inferences. For instance, since the desk is sturdy, I can use it as a step to pick out a book on the upper shelf, but I can't use the laptop for this purpose; if I want to know the time I should look at the clock but not the crow; and so on. These discrete units, which we assume to populate the world, and on the basis of which we think and carry out inferences, have been called *natural kinds* by philosophers. Natural kinds are the basic units not only of our mundane reasoning, but also of sophisticated scientific thinking. Chemists categorize various materials into chemical kinds such as carbon, gold, and argon, and explain chemical reactions in terms of the properties of these elements. Biologists classify all living things into distinct species and study the ecological, behavioral, physiological, and genetic characteristics of each species. These chemical elements and biological species define natural kinds in chemistry and biology, respectively.[15]

The foregoing implies two things. First, "the world" investigated by each scientific field is (or is conceived to be) constituted from those things that are regarded as natural kinds in that particular field. The world according to the chemist is the totality of chemical reactions among various elements, while in the eyes of the biologist, the world appears as a kingdom of all the different kinds of living creatures. Second, the way the world is articulated may well differ from one scientific context to another. From a chemical perspective, I am distinct from you because we clearly differ in the constitution, amount, and ratio of chemical substances that compose our bodies. From a biological perspective, however, you and I (I believe) are both members of the same *Homo sapiens* species, and in this respect we are no different. This is not necessarily to say that the chemist's picture is finer grained and more precise, while biology gives us only a rough and vague sketch. For one thing, some inferences become possible only by abstracting from the finer differences. For instance, if I feel safe taking an approved drug that has passed clinical trials, this is only because I believe I belong to the same human species as the participants of the trials, so that the drug should exercise similar effects on me, including collateral ones. If, in contrast, I am too strict and deny any commonality between myself and the trial participants because all humans differ at the molecular level, I could not hope for any general knowledge about human beings. In this way, natural kinds provide us not only with an articulation of the world, but also with the ontological identity criteria according to which we judge what is the same as or different from something else, thereby underpinning our inferential and explanatory practices both in mundane and scientific contexts.

Now going back to statistics, the preceding discussion suggests that distribution families serve as natural kinds in statistics. For this reason, I propose to call them *probabilistic kinds*. Just as chemists explain various reactions by positing chemical elements, statisticians explain regularities in data by postulating probabilistic kinds. Just as a chemical element is identified by its atomic weight and electron arrangement, a probabilistic kind is characterized by its parameters. And just as chemists study various materials by determining their constituent elements, statisticians solve inductive tasks by reducing them to a particular probabilistic kind. The only difference is that while chemical kinds are fully determined by their physical basis, probabilistic kinds are not necessarily so. As we saw earlier, the choice of an appropriate family of distributions depends not so much on the physical setup of a given inductive problem as it does on its context and the way it is framed. Or, in philosophical jargon, probabilistic kinds *do not supervene on* their physical components.[16] Recall that we modeled a coin toss by a Bernoulli distribution. That was because we supposed the coin will land either heads or tails and ignored the possibility of its landing on its edge. If we are to take the latter possibility into account, we should use a multinomial distribution with three categories. As this example shows, the question of which statistical model we should use is not just determined by the physical nature of the object under study, but rather depends on the modeler's interests and intentions. This, however, by no means prevents probabilistic kinds from being "natural." For, as we have already seen, the choice of natural kinds also reflects the scientist's interests, with the consequence that the world is articulated differently depending on the scientific context and questions being asked. The important point is that probabilistic kinds furnish the basic building blocks of inferential statistics, in terms of which statisticians frame inductive problems and predict the future. Some examples of probabilistic kinds that were introduced in this section include the Bernoulli, binomial, and normal distributions, but it should be emphasized again that this is just a very small sample. Figure 1.4 is a visual map of (some, not all) families of distributions and their mutual relationships (Leemis and McQueston 2008). This may look daunting to some, but its significance may become clearer if you take it as a kind of "periodic table" that exhibits the glorious results of statisticians' attempts to pin down the essences of a variety of inductive problems. Had Mendelejev's periodic table contained

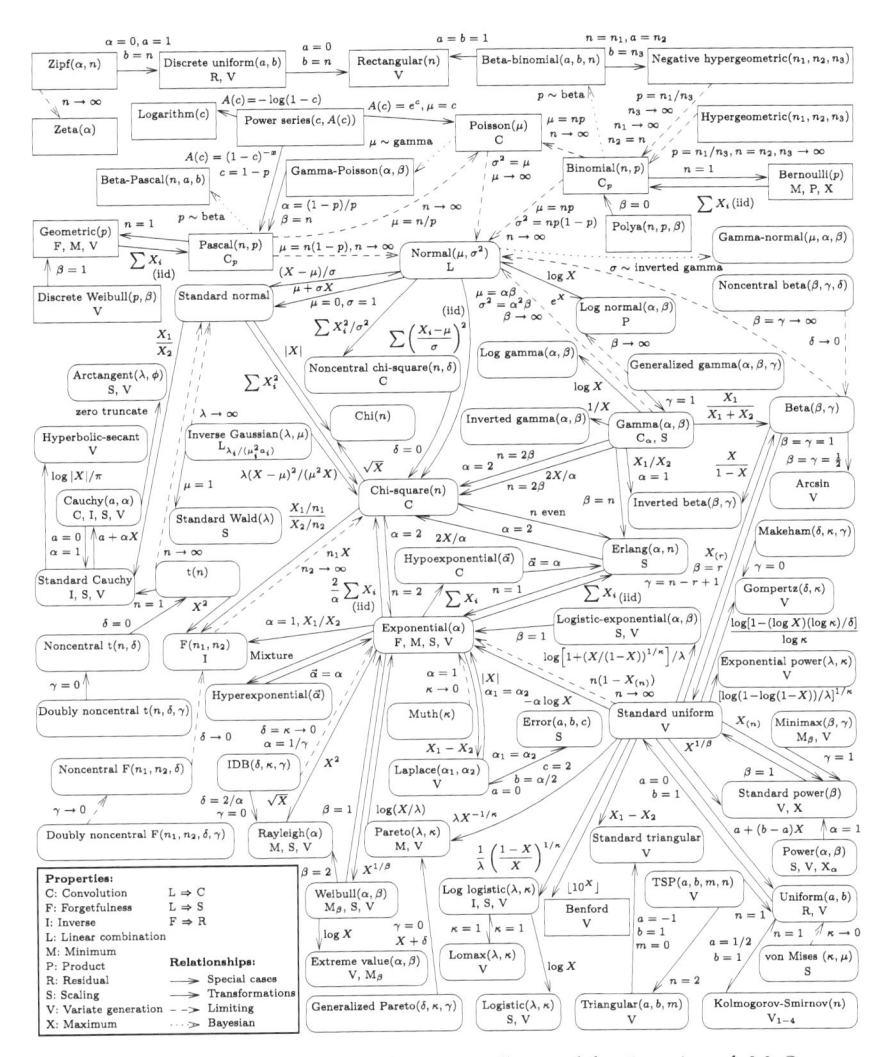

FIGURE 1.4 Various families of distributions as illustrated by Leemis and McQueston (2008). In the upper right you can find the three distribution families discussed in this section. Reprinted by permission of Taylor & Francis Ltd, http://www.tandfonline.com on behalf of American Statistical Association.

only a few atoms, chemistry would have been incomplete and insufficient. Likewise, if we had only the binomial and normal distributions, the study of statistics surely would have been much easier, but as a discipline it would have been meager and devoid of practical applicability.

Let's wrap up. We began this section by noting that to tackle the problem of induction, inferential statistics adopts a dualistic ontology, in which a probability model is posited as a uniform structure behind the data. Parametric statistics further articulates this posited uniformity/probability model in terms of probabilistic kinds, which are represented by specific functional forms (families of distributions). Just as chemists explain reactions in terms of chemical elements, statisticians solve inductive problems by reducing them to appropriate probabilistic kinds.

The world is segmented into natural kinds, and the goal of science is to identify and exploit these kinds for the purpose of inference and explanation—this idea may sound commonsensical, if not a banal truism, to some. However, the claim is not as innocent as it might appear. For one thing, it does not square well with the empiricist philosophy, or in particular with the positivism discussed in the previous section. This is because believing in natural kinds seemingly amounts to postulating latent substances that go beyond individual phenomena—like a "gold itself" that cannot be reduced to any particular instance of gold—and aiming to explain the individual phenomena in terms of these idealistic postulates. We will not go into the philosophical debates over natural kinds here,[17] but we should recall that it was precisely this kind of realist idea—that there are entities that go beyond concrete, tangible phenomena—that was the target of the positivist expulsion spearheaded by Mach and Pearson.

It is not without reason that empiricists try to avoid natural kinds. The assumption of natural kinds behind the data commits one to a richer ontology that postulates more entities than does empiricism or positivism. One cannot, however, just make a postulation and leave. Realists bear the burden of providing an epistemological method for probing such "latent entities" from experience, which is surely a challenging enterprise. In the present statistical context, one must provide a method that allows us to select the appropriate distributional family, examine and justify the choice, and furthermore estimate its parameters for a given inductive task. The essence of inferential statistics consists in this *epistemology*, i.e., the art of making inferences about the underlying probability model on the basis of observed data. This inferential methodology comes in various styles, each forming a distinct statistical tradition such as Bayesian statistics, classical statistics, model selection, and so on. The next three chapters will focus on each of these schools, with particular emphasis on their epistemological character.

Further Reading

There are so many textbooks on statistics. Here I pick just a few: the most concise (Hand 2008) and the more comprehensive (Wasserman 2004; Efron and Hastie 2016). Vaughan (2013) gives a good summary of descriptive statistics, probability theory, random variables, and so on. The historical development of statistics is documented by Stigler (1986), Hacking (1990, 2006), and Salsburg (2001). Salsburg's book is fun and readable, but it has also received some criticism (Porter 2001). Hacking is also known for his seminal work on the philosophy of statistics (Hacking 2016). The *Stanford Encyclopedia of Philosophy* (https://plato.

stanford.edu/index.html) should cover most of the philosophical concepts that appear in this and the following chapters, including positivism, natural kinds, and so on.

Notes

1. It is also common to use $n - 1$ instead of n for the denominator of the sample variance, so that it is an *unbiased* estimator of the population variance (an unbiased estimator is an estimator whose expected value equals the true value of the estimand). This also applies to other statistics such as the sample covariance.
2. This terminology calls for a cautionary remark. In this book we follow the convention in probability theory and define a probability model as a triplet consisting of a sample space, an algebra thereof, and a measure (probability) function. Some statistics textbooks, however, call this the "true distribution" and use the term "probability model" to denote a *model* of this true distribution, or what we will later call a "statistical model." The reason we adopt the probability-theoretic convention is because the assumption that a given inductive problem can be expressed in terms of a certain probability distribution already involves probabilistic modeling (after all, the uniformity of nature is nothing but a model). In this parlance it becomes crucial to sharply distinguish probability models from statistical models, as will be stressed in Section 1.2.3.
3. Reichenbach (2008), in his doctoral thesis originally published in 1916, took a probability model as a theoretical posit that allows for inductive reasoning. I thank Clark Glymour for the pointer.
4. Here the number of the subsets in the antecedent $A_1, A_2, \ldots$ may be infinite (but must be countable). The same is true with the probability axiom A3, described next.
5. One might then wonder: why not use power sets all the time rather than introduce the esoteric σ-algebra? Indeed, power sets will work when the sample space is finite, but not when it is uncountable, as in the real interval. The power set of an uncountably infinite set is too large and rife with pathological subsets whose size cannot be meaningfully measured. This is why we need to limit the range of events to the well-behaved σ-algebra. A set equipped with a σ-algebra is called a *measurable space*, because we can meaningfully measure the size of its parts.
6. Note that this is a function not from Ω but from a σ-algebra defined on Ω to $[0, 1]$, because it assigns a value to *subsets* of Ω.
7. We will use this common notation throughout this book, but it may be confusing at first. The equal sign here differs from its standard usage of asserting the equality of the left- and right-hand sides. Thus, $X = 165$ does not mean "X is 165," but rather "the set of elements for which X gives the value 165."
8. Here Σ_x means the sum over all possible values of X. If X is continuous, the sum is replaced with an integral: $P(y) = \int_{-\infty}^{\infty} P(y, x) dx$. However, we will not distinguish the two cases in this book. The meticulous reader should read the summation symbol as an integral in the case of continuous random variables.
9. A bit more precisely, for data with n samples we consider a sequence of n random variables, and the requirement is that these random variables are independent.
10. Note that not all statistical inferences depend on the IID condition. Samples from a time series or data with a spatial structure, for example, are most likely not

11. This is actually not entirely correct. The central limit theorem assumes that the population mean and variance of the underlying distribution are well-defined and non-divergent. Some distributions such as the Cauchy distribution do not satisfy these conditions.
12. While parameters are sometimes confused with expected values, the two should be distinguished. As we discussed earlier, expected values are properties of a probability model that is unknown but presumed to exist, while the parameters of a distributional family identify a hypothetical and arguably fictional statistical model. They match numerically only when the distributional hypothesis is true, but even in such cases they are conceptually distinct.
13. The discussion here is reminiscent of Nancy Cartwright's *entity realism*. Cartwright (1983) argues that while objects of physics like electrons and quarks exist, the fundamental laws about their properties and behavior are idealized and simplified, and hence, strictly speaking, false (or as she puts it, "lies"). Likewise, one can say that although probability models *qua* uniform objects exist in nature, the statistical models that express their distribution in an idealized and simplified form are "lies" or, to use Box's word, "wrong."
14. Enumerative induction is a type of inductive inference that aims to confirm a general proposition through repeated observations. For instance, concluding that all ravens are black based on numerous observations of black ravens is an example of enumerative induction.
15. There are, however, debates in the philosophical literature regarding the ontological status of chemical elements and biological species, including whether they really constitute natural kinds; see Sober (1980) and Boyd (1999).
16. When one thing or property is fully constituted by other things or properties, the former is said to supervene on the latter. For instance, color is said to supervene on the wavelength of light, and the temperature of a gas on the mean kinetic energy of its particles.
17. A longstanding question that has especially been the subject of extensive debate is whether natural kinds exist objectively on the side of the world, or are theoretical constructs of particular scientific theories—a question that dates back to the medieval problem of universals. In this book I adopt a view closer to the latter, but without attempting to justify it. Interested readers are referred to Bird and Tobin (2018).

2
BAYESIAN STATISTICS

In the previous chapter we introduced the ontological basis of inferential statistics and discussed the respective roles of probability models and statistical models. A probability model embodies the "uniformity of nature" that goes beyond what is observed and serves as a theoretical ground for making inferences about unobserved data. A statistical model or "probabilistic kind" is a further model of this presumed uniformity, expressed in a familiar functional form. Statistical inferences then proceed by making hypotheses about the parameters of these posited models, which are to be estimated and assessed vis-à-vis the observed data. Our next task is to take a closer look at this inferential procedure, and this is where we encounter the first point of divergence. As announced, the interpretation and methodology of statistical inference—namely, what counts as an inference and how it should be carried out—vary among different statistical schools such as Bayesian statistics, classical statistics, and model selection. A particularly notable and historically important difference is the one between Bayesian statistics and so-called classical statistics. Roughly speaking, in Bayesian statistics a probability expresses the degree of certainty of a given hypothesis and is to be updated on the basis of data; whereas in classical statistics a probability expresses the frequency of data, on the basis of which statisticians make a definitive judgment about the hypotheses under consideration. This chapter takes up Bayesian statistics, first giving a brief overview of its methodology and then examining its philosophical implications. Classical statistics will be examined in the next chapter.

2.1 The Semantics of Bayesian Statistics

As mentioned earlier, in Bayesian statistics inductive reasoning is understood as a process of updating the degree of certainty of hypotheses on the basis of data.

But before implementing this idea, a bit of preparation is in order. That is, we need to interpret the mathematical apparatus of probability models introduced in the previous chapter within the actual context of inductive reasoning, so as to make clear *what probability is in the first place*. Both probability models and statistical models are our theoretical constructs, or, to be more precise, mathematical entities defined within a set-theoretic framework. Such mathematical constructions allow us to explore, among other things, the formal properties of probability distributions and the limiting behaviors of certain estimators through rigorous derivations. But in order to apply these mathematical results to concrete problems, we have to interpret the abstract models and make clear what in the real world they represent. Such interpretive work is a prerequisite for any scientific investigation that uses models, not just mathematical ones but of all sorts. If an astronomer wants to use a differential equation model to predict planetary orbits, or if a molecular biologist wants to explain the hereditary mechanism using the double helix model, they must specify what is represented by each component of their models, and to which aspect of their target systems it corresponds. The same holds for probability models. In the previous chapter we defined probability as a function that measures the "size" of an event, or subset of a sample space, in terms of a real number ranging from zero to one. But what does this "size" of a set measured in numbers stand for? An intuitive answer would be that it measures how "likely" an event is to occur. This, however, is far from satisfactory—since events on a sample space are by definition just sets, it hardly makes sense to say that such mathematical entities are "(un)likely to occur." Even if we put this aside, the very meaning of "likelihood" (here we are using this term in the common English sense, as a synonym of chance, not as a technical term to be defined shortly) is not clear at all. When we say that something is likely to occur, are we talking about an objective property that it has by itself, and if so, what kind of property is this? Or if likelihood is a subjective property, who is the subject, and how should it be determined?

Thus, if we are to understand how and why the mathematical apparatus introduced in the previous chapter can ever be applied to real-world inductive problems, we need to clarify the *semantics* of probability, or what probability actually means. This is where the first philosophical conflict emerges. The probability-theoretic ontology introduced in the previous chapter is a common premise of inferential statistics in general. However, there are different interpretations as to its semantics—what in the real world these mathematical posits represent—which in turn form the bases of the two major epistemological approaches in inductive inference, namely Bayesian and classical statistics. Roughly speaking, Bayesian statistics usually interprets probability as a subjective degree of belief, while classical statistics takes it as the objective frequency of an event's occurrence. For this reason, the former has traditionally been called subjectivism (or more commonly Bayesianism, which is taken to imply the subjectivity of probability), and the latter frequentism.

These labels capture a contrast between the two approaches in one respect, but they are misleading in another. In particular, they may blur the important distinction between semantics and epistemology. While the questions of what probability is and how it should be interpreted are semantic in nature, the main disagreement between Bayesian statistics and classical statistics is epistemological and concerns the proper methodology of inductive inference. These questions—semantic and epistemological—are logically distinct, so there is no contradiction in, say, interpreting probability values in classical statistics in a subjectivist fashion. That being said, it is still true that semantics (subjective/objective) and epistemology (Bayes/classical) are deeply intertwined. For this reason, in this and the following chapters, we introduce the subjective and frequentist interpretations of probability as the semantics for Bayesian and classical statistics, respectively. But the reader should bear in mind the conceptual distinction between semantics and epistemology.

With this in mind, let's begin with the subjective interpretation of probability. First, we attach a meaning to the sample space. The sample space for Bayesians is a collection of various propositions. For instance, the sample space of rolling a die once, $\Omega = \{1, 2, \ldots, 6\}$, consists of propositions such as "the 1-face is up," "the 2-face is up," ... "the 6-face is up." Let us denote the proposition "the i-face is up" by A_i. Events are formed by attaching logical connectives to these atomic propositions. The event of getting an even number, say, is the composite proposition $A_2 \vee A_4 \vee A_6$, while that of getting a number other than 1 is $\neg A_1$. We expect these events to satisfy the rules we saw in the previous chapter. Restated in the framework of propositional logic, those rules become:

R1 A contradiction[1] $\bot$ is an event.
R2 If proposition A is an event, so is its negation $\neg A$.
R3 If propositions $A_1, A_2, \ldots$ are events, so is their disjunction $A_1 \vee A_2 \vee \ldots$ (A_1 or A_2 or $\ldots$).

That is, a set of composite propositions that qualify as events are closed under negation and disjunction (and hence, under conjunction, too). From R1 and R2, a logical truth or tautology $\top$ also counts as an event and corresponds to the total event (Ω itself). A set of events that satisfy the above rules is called a *Boolean algebra*. The algebra of a probability model in the subjective interpretation of probability, therefore, is a Boolean algebra.

A probability function is then a function that assigns a number from zero to one to a proposition-event in such an algebra. Under the subjective interpretation, a number given by such a function represents *the degree of belief* of an epistemic agent toward the corresponding proposition. We will come back to this concept shortly; for now, the degree of belief that a certain person has in a proposition can be understood as the strength with which that person believes in the truth of the proposition. The probability axioms under this interpretation are:

A1 $0 \le P(A) \le 1$ for any proposition A.
A2 $P(T) = 1$.
A3 If propositions $A_1, \ldots, A_n$ are not consistent, so that $A_i \wedge A_j \iff \bot$ for $1 \le i < j \le n$, then

$$P(A_1 \vee A_2 \vee \ldots) = P(A_1) + P(A_2) + \ldots.$$

A1 says that any degree of belief lies between zero and one. According to A2, a logical truth has the maximum degree 1, which indicates absolute certainty. Finally, A3 states that the degree of belief for the possibility that at least one of a mutually inconsistent set of propositions is true must be the sum of the degrees of belief for each proposition.[2] Under this interpretation, the axioms of probability prescribe the rules that any epistemic agent engaging in inductive reasoning must follow in calibrating his or her belief in any proposition.

But who in the world is this "epistemic agent engaging in inductive reasoning"? That is, whose degree of belief does probability represent? Is it a particular individual conducting inferences, such as you or me? Does a group of people with common interests or issues such as a scientific community or political institution also qualify as an epistemic agent? Or is an epistemic agent only an ideal agent that need not actually exist? For the time being, let us be lenient and admit all of these as epistemic agents. We also allow the degree of belief in the same proposition to differ from person to person. I may believe firmly in the existence of intelligent extraterrestrial life, whereas you might not. Groups of scientists such as NASA or SETI might have different takes. Each epistemic agent may assign probabilities differently and thus have their own probability function P. For the meantime, let's not question which probability function is correct. The probability axioms do not prohibit you from believing strongly in a wildly weird idea, like "the moon is made of blue cheese." They only impose general and abstract rules on degrees of belief (such as that the degree of belief in an inconsistent proposition must be zero); they do not concern which specific propositions are believed or disbelieved by any particular agent.[3]

Nonetheless, we still need to consider how such degrees of belief are measured, and why they must conform to the aforementioned axioms. These are major topics in the philosophy of probability, which we will touch upon only briefly in this book; the interested reader is referred to the relevant literature (Gillies 2000; Childers 2013; Rowbottom 2015). The canonical way to measure degrees of belief appeals to the notion of fair odds. For any proposition A, consider a lottery such that you win $10 if A turns out to be true, and nothing otherwise (i.e., if $\neg A$ holds). The degree of belief that a given epistemic agent—let's say it's you—has in this proposition A can be determined by what you deem to be the fair price of this lottery. Here the "fair price" is the price at which you are willing to either buy or sell the lottery ticket. Thus, if you evaluate it as $6, you must be willing to buy it from me for $6 (in that case you will win $10 from me should A be true), and conversely, you have to sell it to me for the same price if I ask for it (in that case you will have to pay $10 to me should A be true). Under this arrangement, what is the fair price of the ticket? Clearly, that should depend on how strongly you believe in A (and perhaps on many other things, such as whether you are into gambling, and how appealing you find the $10 prize, but let us ignore these matters here). If A stands for a logical truth such as "It will either rain or not rain in Kyoto on the next New Year's Day," you might feel safe to pay up to $10. In contrast, you may not want to pay that much if the proposition is "It will rain in Kyoto on the next New Year's Day." Again, if the proposition is "It will snow in Hawaii on the next New Year's Day," you would probably find the ticket almost worthless. These examples suggest that the value a person attaches to a bet on a given proposition reflects that person's degree of belief in that proposition. Proponents of the subjective interpretation of probability thus argue that the probability value of that belief can be determined by the ratio of the named price to the prize. If, for instance, you think the fair price of a lottery that will win you $10 if A holds is $3, the probability you assign to A will be 0.3.

Once degrees of belief are defined in this way, it becomes clear why they must conform to the aforementioned axioms of probability. Let A again be "It will rain in Kyoto on the next New Year's Day," with its negation $\neg A$ being "It will not rain in Kyoto on the next New Year's Day." Suppose you assign a probability of 0.6 to both of A, $\neg A$, in violation of the probability axiom. This means that you are willing to either buy or sell for $6 the lottery ticket that will win you $10 when A holds and the ticket that will win you $10 when $\neg A$ holds. I thus sell you both tickets for the total price of $12. But come what may, your net gain will only be $10, for only one of A or $\neg A$, but not both, can be true simultaneously. You are thus destined to lose $2. Conversely, now suppose you assign a probability of 0.4 to both of A, $\neg A$. I then buy both for the sum of $8. Since one of them will certainly be a winning ticket, I get $10 and you lose $2 again. A series of bets that is sure to incur a loss as in this example is called a *Dutch book*. To avoid being Dutch-booked in the example, your probability assignments must satisfy $P(A) + P(\neg A) = 1$. The preceding discussion illustrates that a violation of the second axiom of probability—where one assigns a probability other than 1 to a logical truth such as $A \lor \neg A$—incurs a sure loss.

It can be easily shown that a violation of the other two axioms also leads to similar Dutch books. Now, a person who can be "Dutch-booked" in this way—i.e., a person who readily agrees upon odds that imply a certain loss—is surely not rational. Hence, if we want to be rational epistemic agents, we must assign our degrees of belief in accordance with the axioms of probability, and conversely, the degrees of belief of any rational agent will satisfy the probability axioms—or so argue the proponents of the subjective interpretation.

The subjectivist idea of reducing probability values to a person's beliefs and measuring them in terms of bets may strike some as being too arbitrary and unscientific. This approach, however, does have its own advantages. One is its versatility: anything expressible in a proposition can be assigned a probability value. A case in point is a hypothesis about the parameters of a distribution. For instance, the hypothesis that a coin is fair can be expressed by a proposition about the parameter of a Bernoulli distribution $\theta = 0.5$, and the degree of certainty of this proposition can be meaningfully represented by a probability value. This is in stark contrast to the frequentist framework, where, as we will see in the next chapter, probability values cannot be meaningfully assigned to statistical hypotheses. Moreover, we can update the degree of certainty of the hypothesis by calculating the probability of the hypothesis/proposition for each parameter on the basis of the results of coin-tossing experiments. This updating procedure is guided by the famous Bayes' theorem. In the next section we will look into the basic ideas of Bayesian statistics, which is built upon Bayes' theorem, along with some examples.

2.2 Bayesian Inference

As stated previously, inductive inference according to Bayesian statistics is the process of updating the degree of certainty of a hypothesis on the basis of observed evidence. If we let h be a hypothesis and e be data or evidence, the degree of belief in the hypothesis after the evidence is obtained is represented by the conditional probability $P(h|e)$. By Bayes' theorem, introduced in the previous chapter, this becomes

$$P(h|e) = \frac{P(e|h)P(h)}{P(e)},$$

where

- $P(e|h)$ is the *likelihood*, which represents how likely it is that we obtain evidence e if the hypothesis h is true.
- $P(h)$ is the *prior probability* (or simply "prior"), which represents the degree of certainty of the hypothesis before the evidence is obtained.
- $P(h|e)$ is the *posterior probability* (or "posterior"), which represents the degree of certainty of the hypothesis after the evidence is obtained.

When there are multiple hypotheses $h_1, h_2, \ldots$, their posterior probabilities $P(h_i|e)$ can be calculated in a similar way. By applying the law of total probability (see Section 1.2.1), the $P(e)$ in the denominator can be replaced by the sum of the products of the likelihood and prior probability of each hypothesis:

$$P(e) = \sum_i P(e|h_i) P(h_i).$$

Hence, Bayes' theorem essentially allows us to calculate the posterior probabilities of hypotheses from their likelihood and prior probability; in other words, it gives us a rule for updating our belief in each hypothesis on the basis of its explanatory power (i.e., its ability to predict the evidence that is actually obtained) and our pre-observational belief in that hypothesis.

Bayesian statistics makes full use of this theorem in making inductive inferences such as hypothesis confirmation and prediction. Let's look at some examples.

2.2.1 Confirmation and Disconfirmation of Hypotheses

Imagine that your nearby shopping mall holds a lottery every weekend, where you draw a ticket from a box. They have two kinds of boxes: one of them, say A, has winning tickets in a ratio of 1 out of 10, while the other, B, has winning tickets in a ratio of 3 out of 10. Each weekend they choose one of these two boxes, without letting you know which. On some Sunday you visit the mall and draw a ticket, which, unfortunately, is blank. Given this evidence e, how should you update the probability of the hypothesis h_A that the box in front of you is A?

- Since you had no information as to which box was in place before the draw, it would be reasonable to set the prior probabilities for the two boxes equal, so that $P(h_A) = P(h_B) = 0.5$.
- Since the probability of getting a blank ticket from box A is 90%, we have $P(e|h_A) = 0.9$; likewise, the probability of getting a blank ticket from box B is $P(e|h_B) = 0.7$.

Plugging these into Bayes' theorem, the posterior probability of h_A is

$$P(h_A|e) = \frac{0.9 \times 0.5}{0.9 \times 0.5 + 0.7 \times 0.5} = \frac{0.9}{1.6} \sim 0.56.$$

On the other hand, the posterior of the hypothesis h_B that the box is actually B is

$$P(h_B|e) = 1 - P(h_A|e) \sim 0.44.$$

From this we see that, given the evidence e that you got a blank ticket, the degree of certainty of the hypothesis that the box in front of you is A increases from 50% to about 56%, while the hypothesis that the box is B decreases to about 44%.

All this can also be expressed using random variables, which we introduced in Section 1.2.2. Recall that random variables are functions that express properties or possibilities in terms of their values. The property we are interested in here is the proportion of winning tickets in the box in front of you. If we use the random variable θ to denote the proportion of winning tickets, the hypotheses h_A and h_B can be expressed as $\theta = 0.1$ and $\theta = 0.3$, respectively. That is, θ is a discrete variable with two values {0.1, 0.3}. The prior probabilities of these possibilities or values together define a *prior distribution* of θ, which in the aforementioned case is the uniform distribution that assigns the same probability to both values (Section 1.2.3). Outcomes of the draw can also be expressed by a random variable E, where $E = 0$ represents drawing a blank ticket and $E = 1$ represents drawing a winning ticket. With this notation, the posterior probabilities of the hypotheses given the evidence of a losing ticket can be expressed as $P(\theta | E = 0)$, which defines the *posterior distribution* of θ. Bayesian updating can thus be thought of as a process of calculating the posterior distribution from a prior distribution.

Now, the random variable θ in the preceding discussion represents the parameter of a Bernoulli distribution, which we saw in Chapter 1. This observation allows us to see the Bayesian inference as a process of refining our belief in a probabilistic kind. This proceeds as follows. First, the stochastic process of drawing a ticket from a box is modeled using a probabilistic kind, in this case the Bernoulli distribution. Next, we make hypotheses (two in our example) about the parameters that determine the behavior of this probabilistic kind. Then we update the probabilities of these hypotheses—our degrees of belief in these hypotheses—on the basis of the observed data. Through this process, our belief about this specific instance of the probabilistic kind—the lottery—gets refined. Thus, in the Bayesian formulation, an inductive inference is understood as the process of evaluating hypotheses about a presupposed probabilistic kind in relation to observed data, by updating its distribution from the prior to the posterior.

2.2.2 Infinite Hypotheses

In our example in Section 2.2.1, we considered just two hypotheses about the parameter of a probabilistic kind (the Bernoulli distribution), and calculated their posterior probabilities. This reflects our assumption that there are just two possible boxes with a fixed winning ratio.

As the next step, imagine a more general case where we have no idea at all about the proportion of winning tickets in the box. Our task is, after a certain number of draws, to estimate this proportion, and hence the probability of drawing a winning ticket. If the box contains a sufficiently large number of tickets, we may pretend that the true proportion could be any number between zero and one. This means that there are uncountably many hypotheses to consider for the parameter of a Bernoulli distribution. The random variable θ that represents each hypothesis is thus a continuous variable, and our task is to calculate its posterior distribution. Now suppose we draw n tickets, x out of which are winning (we assume that tickets are replaced after each draw). According to the binomial distribution seen in the previous chapter, the probability of this event under hypothesis θ, i.e., the likelihood of this event, is

$$P(x|\theta) = {}_nC_x \theta^x (1-\theta)^{n-x}.$$

Since by assumption we have no prior information about what is inside the box, the prior distribution $P(\theta)$ should be uniform over all values of θ. Then, by Bayes' theorem, the posterior distribution is

$$P(\theta|x) = \frac{{}_nC_x \theta^x (1-\theta)^{n-x}}{P(x)} \cdot P(\theta).$$

Noting that ${}_nC_x$, $P(x)$, and $P(\theta)$ do not depend on the hypothesis θ, the posterior probability is proportional to $\theta^x(1-\theta)^{n-x}$, that is,

$$P(\theta|x) \propto \theta^x (1-\theta)^{n-x}.$$

Substituting a value from 0 to 1 into θ yields the posterior probability of any parameter hypothesis.[4] Figure 2.1 shows plots of the posterior probabilities of all the hypotheses ranging $0 \leq \theta \leq 1$, calculated from various datasets (n, x). The plots illustrate that as the number of trials increases, a particular range of

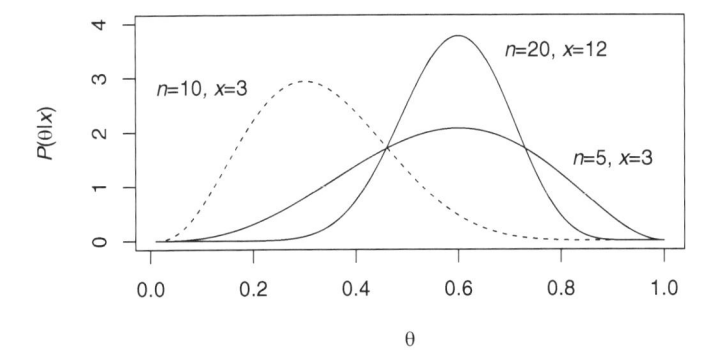

FIGURE 2.1 The posterior distribution $P(\theta|x)$ of the parameter θ of a Bernoulli distribution given x out of n successes. The prior distribution is set to be uniform over [0, 1]. Two solid curves labeled $n = 5$ and $n = 20$ denote experiments with the same success rates, but the inference in the latter is more refined thanks to the increased number of trials.

hypotheses comes to stand out, and the range itself shrinks, leading to a more precise inference. In fact, this posterior distribution itself represents a probabilistic kind, called the beta distribution. In general, the posterior distribution of the success probability θ given x successes out of n trials is represented by a beta distribution with two parameters, $(x + 1, n - x + 1)$.

To wrap up, a Bayesian inference begins by modeling a given problem with an appropriate probability distribution or probabilistic kind. This allows one to form hypotheses about the probability model, i.e., the uniformity of nature, in terms of the parameter(s) of this presupposed distribution, and to calculate their likelihood with respect to the observed data. From this likelihood and prior probability of the parameter hypotheses, Bayes' theorem derives posterior probabilities, thereby updating our beliefs about the probability model.

We have illustrated this process using a Bernoulli model of a lottery, but note that the same Bernoulli model may well be applied to many other random binary problems, such as a coin toss or predicting the sex of a newborn baby. In this sense the Bernoulli model points to a general "type" shared by various inductive problems, whence we have called it a probabilistic kind. Needless to say, the Bernoulli distribution is just one example of such kinds, and other inductive setups may be best captured by different kinds. An inference about the mean height or weight of a particular human population, for example, would be better served by the normal distribution; whereas if we want to estimate the number of occurrence of a certain sporadic event, such as the number of phone calls that happen in a day, or the time interval from one call to the next, the Poisson or exponential distributions would be appropriate. In any case, stipulating a probabilistic kind determines the likelihood of a parameter hypothesis, which, combined with a prior distribution, yields a posterior distribution over the hypotheses. Through such a process, Bayesian epistemic agents refine their beliefs about the hidden uniformity of nature.

2.2.3 Predictions

How, then, do such inferences about a probability model help us in predicting future or unobserved events? As we saw in Chapter 1, a prediction by inferential statistics is mediated by an inference about a probability model. Bayesian statistics implements this general strategy by deriving predictions from an updated posterior distribution. Let us illustrate this using the lottery example from the previous section. The posterior probabilities of the Box A and Box B hypotheses after drawing one blank ticket were $P(h_A|e) = 9/16$ and $P(h_B|e) = 7/16$, respectively. Assuming that we replace the ticket after each draw, we want to predict the outcome of a second draw. Letting $\tilde{e}$ represent the event that in the second draw we get a blank ticket again, the desired probability is $P(\tilde{e}|e)$, read as "the probability of drawing a blank ticket given the evidence e that the first ticket was blank." To calculate this, we first observe that there are two mutually exclusive scenarios in which we draw a blank ticket:

1. The box is actually A, from which a blank ticket is drawn.
2. The box is actually B, from which a blank ticket is drawn.

Let us begin with the first scenario. Having drawn a blank ticket in the first draw, we now believe that the box is A with the probability $P(h_A|e)$, which is the posterior of h_A. The probability of drawing a second blank ticket from this box is the likelihood $P(\tilde{e}|h_A)$. Multiplying them gives the probability of the first scenario as $P(\tilde{e}|h_A)P(h_A|e)$. Since the same goes for the second scenario, the desired probability is given by their sum:

$$P(\tilde{e}|e) = P(\tilde{e}|h_A)P(h_A|e) + P(\tilde{e}|h_B)P(h_B|e).$$

The probability of winning on the second try can be calculated in the same fashion. Together they give the distribution on the second draw informed by the outcome of the first draw, which is called a *posterior predictive distribution*.

While there are only two hypotheses h_A, h_B in the above example, a case involving more hypotheses can be handled in the same way. When there are uncountably many hypotheses as in Section 2.2.2, the posterior predictive probability of $\tilde{e}$ given data e is

$$P(\tilde{e}|e) = \int P(\tilde{e}|\theta)P(\theta|e)d\theta.$$

Here again, the posterior probability $P(\theta|e)$ is our updated degree of belief in a parameter hypothesis θ, while the likelihood $P(\tilde{e}|\theta)$ represents the process of sampling new data $\tilde{e}$ under this hypothesis. The prediction is given by summing up (integrating) the likelihoods of all parameter hypotheses weighted by their posteriors. Note that these two components of Bayesian prediction correspond neatly to the two arrows in Figure 1.2; the posterior probability captures the upward inference to the model from observed data, while the likelihood represents the downward sampling of unobserved data from the estimated model.[5]

2.3 Philosophy of Bayesian Statistics

2.3.1 Bayesian Statistics as Inductive Logic

What was covered previously in this chapter is just the bare minimum of Bayesian statistics, upon which a standard statistics textbook would go on to develop more practical applications and sophisticated techniques. But this is a philosophy book, and we are interested not so much in practical solutions as in epistemological significance. That is, we are interested more in what all these calculations mean, and how they serve our inferential practices. To approach these questions, we need to take a step back and examine the philosophical basis of Bayesian statistics.

According to Howson and Urbach (2006), who are leading proponents of modern Bayesianism, the Bayesian probability calculus provides a rule for inductive inferences, or a sort of *inductive logic*. But what do they mean by this? To understand their claim, let us set aside the inductive part for a moment and briefly recall what "logic" is. When philosophers speak of logic, they usually mean deductive logic, which is a formal system of deriving conclusions from premises in accordance with certain logical rules. Such derivations must be valid, so that the truth of the premises guarantees the truth of the conclusions. The validity of an inference can be understood in terms of the satisfiability of logical formulae. Consider a simple case of deductive inference in which a conclusion B is derived from premises A and $A \supset B$. We can verify the validity of this inference by checking whether the negation of the conclusion $\neg B$ is consistent with the set of premises. This can be done using a truth function V that assigns the truth value 1 to true propositions and 0 to false ones. With this function, the truth of the premises can be expressed as $V(A) = V(A \supset B) = 1$, while the negation of the conclusion can be expressed as $V(B) = 0$. The question, then, is whether there exists a truth function V that simultaneously satisfies the following system of equations:

$$V(A) = 1, \tag{2.1}$$

$$V(A \supset B) = 1, \tag{2.2}$$

$$V(B) = 0. \tag{2.3}$$

Evidently, no such function exists: from the truth condition (or truth table) of conditional "$\supset$," $A \supset B$ is true exactly when A is false or B is true, so Equation (2.2) can be expanded as $V(A) = 0$ or $V(B) = 1$, which contradict Equations (2.1) and (2.3), respectively. Each truth function represents a possible situation by specifying the truth value of propositions. Hence, the unsatisfiability of this system of equations means that there is no situation where the premises are true but the conclusion is false, which amounts to saying that the aforementioned inference is valid.

Bayesian inference, according to Howson and Urbach, can be understood in the same way. Bayes' theorem is a logical rule that derives posterior probabilities as a conclusion from the likelihood and prior probabilities as premises, in the same way that we derived the two deductive consequences of the formula (Equation 2.2) in our logical inference. The only difference is that instead of a binary truth function V, Bayesian inference uses a probability function P that can take any value between zero and one. This, however, does not amount to any difference in their nature as logical inferences. Indeed, the validity of the probabilistic calculus is warranted by the same kind of satisfiability argument: If one deviates from Bayes' theorem and miscalculates the conditional probability

$P(h\,|\,e)$, the conclusion becomes inconsistent with the premises, in the sense that the likelihood, prior, and posterior probabilities they assign cannot be satisfied simultaneously by any probability function. Hence, if we want to carry out inferences consistently with the degrees of belief that we presuppose, we must follow Bayes' theorem.

Let's sum up. Valid deductive inferences are those that assign a truth value to the conclusion consistently with the truth value assignments of the premises, and this is achieved by following sound inference rules. On the other hand, in order for an inductive inference to be valid, it must adjust the degree of belief of the conclusion, represented by the posterior probability, consistently with the degrees of belief of the premises (the prior and likelihood), and this is achieved by following Bayes' theorem and the axioms of probability. It is in this sense that the Bayesian probability calculus provides us with a logic of inductive reasoning.

2.3.2 Bayesian Statistics as Internalist Epistemology

Let us grant Howson and Urbach's picture of the Bayesian probability calculus as an inductive logic that assigns probability values to propositions in a consistent fashion, in the same way that one assigns coherent truth values in deductive logic. In this picture, Bayesian agents employ the probability axioms and theorems to adjust their degrees of belief about a presupposed probabilistic kind in accordance with prior probabilities and observed data. But pursuing this analogy raises questions as to the alleged *inductive* nature of the inference. For one thing, all that deductive logic does is unpack what is already contained in the premises, without adding any new information. When we take "All men are mortal" and "Socrates is a man" as premises and conclude "Socrates is mortal," we can by no means pretend to have gained any new knowledge. As we saw in the previous chapter, however, the essence of inductive reasoning lies in the fact that it tries to go beyond this limitation of deduction and derive some new information not contained in the premises. But if Bayesian statistics, just like deductive logic, is all about checking the consistency of probabilistic premises through transformations of formulas, how could it be of service to such ampliative inference? In other words, insofar as Bayesian statistics is part of a deductive mathematical theory, how could it be an *inductive* logic at all?

This question brings us to the realm of philosophical epistemology. As we all know, logic and probability theory provide powerful tools for organizing and exploring what we believe and what we know. Under their guidance, we can discover unintuitive mathematical truths or find unexpected inconsistencies among seemingly benign statements. All of them, however, are *a priori* relationships among given propositions. How can they have anything to do with acquiring or inferring *a posteriori* knowledge, as we aim to do when we try to (dis)confirm a hypothesis or predict the future? Philosophical epistemology has been concerned with this very question, trying to understand how our experience and logic relate to knowledge, and in particular how the former can secure the latter. Insofar as Bayesian statistics is an art of producing scientific knowledge through its "inductive" logic and observed data, it must have its own stance on this epistemological question. What, then, does this epistemology look like, and in what sense are the conclusions of Bayesian statistics entitled to the claim of knowledge? With a view towards answering these questions, the rest of this chapter aims to characterize Bayesianism as a kind of epistemological theory.

Epistemology and the Problem of Justification

The key to bridging statistics and epistemology is the concept of *justification* (Nagel 2014; Steup and Neta 2020). To see this connection and understand what justification is, let us leave statistics for a moment and step into the land of epistemology. Philosophical epistemology has long been concerned with the nature of *episteme*, or what we now call knowledge (Pritchard 2014). What is knowledge? The traditional standard answer ever since Plato is that knowledge is a *justified true belief*. First, knowledge must be a belief: if some person is said to know that *P*, she or he must definitely believe that *P*. Moreover, it must be true—that is, it must indeed be the case—that *P*; one cannot "know" something that is actually false. A merely true belief, however, does not automatically count as a piece of knowledge. This is because our beliefs may sometimes happen to be true by sheer coincidence. Suppose that one day you had a hunch that you were going to win a lottery out of the blue, and the ticket you bought was actually a jackpot ticket. You can hardly claim in this case, however, that you *knew* that the ticket you were going to buy was a jackpot, since by assumption you had no reason to think so. Alternatively, imagine that a devout political partisan, who had a deep-rooted desire for the downfall of the opposing party, was led by this desire to believe that its leader is involved in corruption. Imagine further that this leader later faced prosecution and pleaded guilty. But even after this incident, one can't say that the devout partisan indeed knew about the corruption. For in this case, although she may have had a strong *motivation* for so believing, such wishful thinking does not count as a valid *reason* for the belief. Or in other words, her (true, as it turned out) belief lacked the proper justification. For these reasons it has been agreed among philosophers that knowledge is more than just a true belief—it must further be justified by some legitimate reason or evidence.

The purpose of justification is to rule out "lucky guesses" and distinguish genuine knowledge from those beliefs that are true merely by accident (Pritchard 2014). By replacing "beliefs" with "hypotheses," one easily sees that such concerns arise not only in mundane inferences like those mentioned earlier, but also in all sorts of scientific investigations. The question of what counts as scientific knowledge is an intricate problem that cannot be given a simple answer (at least, not here), but one thing that can be said for sure is that scientific knowledge is not just a hypothesis that turned out to be true. This becomes evident when one takes mathematical knowledge as a first approximation. Suppose I believe that any even number greater than 2 can be expressed as the sum of two primes. Suppose further that one day this proposition—known as Goldbach's conjecture—is proven correct by a talented mathematician. Even then, no one would acknowledge that I *knew* the answer to this age-old mathematical conundrum at the time of 2020. For one thing, although my belief may well have been correct, it lacked the requisite mathematical justification, namely a proof.

The empirical sciences may differ from pure mathematics in many respects, but they are alike in that they both require not only truth but also justification for a hypothesis to count as knowledge. If truth were the only criterion, science would become prophecy: for example, one could argue that Democritus knew the atomic theory in that he claimed that all matter is composed of atoms, or that Leibniz knew the theory of relativity in that he denied Newton's absolute space and time. Just as mathematical knowledge requires proof, what makes scientific knowledge *knowledge* is the fact that it is justified by certain procedures and inferences. For this reason, scientists are meticulous about the methods they use in their experiments and observations, as well as the logic that connects the obtained data with their conclusions, and they make them explicit in the "materials and methods" section of their paper. Among this array of justificatory methods, statistics plays the central or even paramount role in the contemporary sciences. Most if not all scientific hypotheses are stochastic, which means that no amount of observation or rigorous experiment can rule out the possibility that any outcomes favorable to the hypothesis under consideration were obtained just by sheer chance. Scientists, therefore, bear the burden of showing that their experimental results or observations are not a mere fluke, but genuine evidence that can indeed support or justify their hypothesis. Statistics takes on this justificatory task in contemporary scientific reasoning.

This perspective suggests that we view the Bayesian calculation of posterior probabilities from observed data as a process of justifying hypotheses. But if so, in what sense is it a justification process? There are a variety of different conceptions of justification, and contemporary epistemologists have discussed their relative merits and demerits (Fumerton 2002; Pappas 2017). What does it mean to justify a belief or hypothesis, and how is it achieved? From one standpoint, justified beliefs are those that are derived via valid inferences from other beliefs that have already been justified. Another position emphasizes the role of a certain objective process in justifying beliefs, rather than the internal relationships among beliefs. Roughly speaking, the former is called internalist epistemology, the latter externalist epistemology. In what follows I argue that, in view of these contrasting philosophical theses, Bayesian statistics is more akin to internalist epistemology. Of course, this does not mean that all internalist epistemologists are clandestine Bayesians, or that Bayesian statisticians are tacit internalists. Rather, my main goal is to show that Bayesianism and internalism adopt similar concepts of justification and also face similar difficulties, and that through such a comparison we can shed light on the epistemological character of Bayesian statistics. With this objective in mind, let us first look at the philosophical thesis of internalist epistemology.

Internalism

What condition does a belief need to satisfy in order for it to be counted as knowledge? According to *internalist epistemology*, an important condition is that the subject who entertains the given belief also possesses a reason or evidence for that belief. Consider again the devout party supporter. Recall that we judged that she did not know the corruption of the opposing party's leader because her conviction came entirely from wishful thinking rather than from a legitimate reason or evidence. But suppose instead that she actually had a decent reason. For example, she may have been a skilled journalist and received a reliable leak to the effect that the FBI was probing the case. Under this new circumstance, we may judge that she had indeed known the corruption before it became public. For in this case, she had possessed a legitimate ground for believing that the party leader is corrupt, namely, the information she got about the FBI's investigation. Here, the ground supporting the given belief needs to be not only veridical, so that the FBI is actually conducting the investigation, but also entertained by the subject in question, so that the reason is "internalized" as her belief—otherwise we cannot say that it was actually her, rather than someone else, who knew that matter. In this way, internalists understand justification as a certain kind of relationship that holds among beliefs possessed by a subject. A belief, according to this view, is justified only if it is related to other beliefs of the epistemic agent that serve as reasons for the target belief.

Such a justificatory relationship can be understood as an inferential one. The leak from the FBI supports the belief in corruption only if the latter can be safely derived from the former. In general, a belief is warranted when it follows from other beliefs of the epistemic agent via an inference based on valid inferential rules. Paradigmatic examples of such inferential rules are those of deductive logic. If I believe that Socrates is a man and that all men are mortal, then an application of syllogism justifies my also believing that Socrates is mortal.

Not all inferences, however, possess this kind of logical necessity. In the example, the tip about the FBI's move may well provide partial evidence for the party leader's crime, but it will not entail it with absolute certainty. In this regard, Bayes' theorem seems to furnish this kind of nondeductive reasoning with a solid inferential rule. Recall that for Bayesians, a probability measures a degree of belief, and that statistical inferences concern relationships among beliefs. Bayesian agents apply Bayes' theorem to premises consisting of the evidence, prior probabilities, and likelihood in order to derive their new degree of belief in a hypothesis. We have seen earlier that this procedure takes the form of a valid inference based on logical rules. Of course, we should not forget that what is concluded through this kind of Bayesian inference is not the truth or falsity of a hypothesis, but rather its certainty as measured by the posterior probability. Hence, what is justified by a posterior probability of, say, 0.99 is not the belief that the hypothesis is true, but rather that it is almost certain. Indeed, there is a very significant gap between these seemingly similar beliefs.[6] But apart from this difference, one may well say that Bayesian statistics embodies, like deductive logic, inferential processes that warrant one's belief in a hypothesis on the basis of one's beliefs in the premises.

We can thus give an answer, from the Bayesian perspective, to the question we posed earlier: how and in what sense does statistical analysis justify scientific hypotheses? According to internalist epistemology, to justify a belief is to support it by other beliefs—which play the role of premises—via valid inferential rules. Bayesian statistics gives substance to this internalist conception of the justificatory process by providing an inferential rule that derives the posterior degree of belief in a given hypothesis from the prior (degree of) belief in that hypothesis, observed evidence, and likelihood. By applying this rule, Bayesians can justify the degree of certainty they attach to a particular hypothesis in light of data and prior knowledge. Bayesian justification, therefore, essentially consists in the inferential relationships among the beliefs possessed by an epistemic subject, and in this sense it can be characterized as internalist.

2.3.3 Problems with Internalist Epistemology

For the sake of argument, let us agree with internalists that justification consists in deriving (the degree of) one belief from (the degrees of) other beliefs in a consistent way. This, however, is just a definition. We need to take a step further and ask whether the justification concept thus defined satisfactorily fulfills the roles it is expected to serve. The primary role of justification, as noted previously, is to prevent "lucky guesses." As epistemic agents, we seek truths, but whether or not a given inductive hypothesis is true is something that lies outside of our purview and is ultimately for the world to decide; one could even argue that we have no way of checking or observing the veracity of our hypotheses. But this doesn't mean that all is lost. There may be sufficient reason to consider certain beliefs that are obtained through a particular method to be true, even if only God knows whether they are ultimately so. To prevent "lucky guesses" is to grant this kind of reason to certain beliefs or hypotheses. We therefore expect justification to warrant, at least to some extent, the truth of the justified belief. This does not mean, of course, that justification must endow beliefs with absolute certainty. Although in the past there were philosophers who, like Descartes, sought an infallible criterion of truth, their contemporary descendants are much more modest and accept that the conclusions of ideal and seemingly flawless reasoning may nevertheless turn out to be false. Still, the concept of justification is expected to be *truth-conducive*, meaning that it must serve as a guide toward the truth, perhaps not infallible but nonetheless reliable to a certain degree (Goldman 2009; Pritchard 2014). The question, then, is this: is the internalist conception of justification we summarized in this section truth-conducive, and if so, in what sense?

The reason this raises a difficulty for internalists is that it is not clear at all how the justification of a belief in terms of its logical or inferential relationships with the agent's other beliefs should contribute to its truth, which is usually understood as a correspondence with the world "external" to the agent. If, for example, I believe that the moon is made of blue cheese and that blue cheese is tasty, I can logically deduce that the moon is tasty; but it is certainly absurd to actually believe in this conclusion. Obviously, the problem here is in the falsity of the premise that the moon is made of blue cheese. Logical deduction warrants the truth of the conclusion only if its premises are also true. What, then, warrants the truth of the premises? If we are to remain within the internalist framework, the necessary warrant cannot come from anything other than further justification, namely, from valid derivations of the premises in question from other beliefs. One can then easily imagine those beliefs in turn requiring yet other beliefs for their own justification, leading to an infinite regress. We finite beings, however, cannot complete such an infinite chain of justification. Historically, this *regress problem* has presented itself as the first obstacle for internalists in establishing the truth-conduciveness of their concept of justification (Hasan and Fumerton 2018).

The same kind of problem may also arise for Bayesianism *qua* internalist epistemology. As a piece of mathematical theory, Bayes' theorem allows us to adjust our degree of belief in a hypothesis given certain evidence in accordance with the likelihood and the prior probability. A coherent assignment of probabilities to beliefs, however, falls far short of vindicating the correctness of the posterior probability, unless we also have sufficient reason to believe in the legitimacy of the premises. Hence, just as in the case of internalist epistemology, the justification of these premises proves essential for the truth-conduciveness of Bayesian inferences, so that their posterior probabilities correctly reflect the way the world actually is. And, as we will see shortly, we encounter a regress situation here similar to the one that has troubled philosophical epistemology. With this in mind, in what follows we will examine in turn how the two major premises of Bayesian inference, prior probability and likelihood, are justified.

Justification of Priors, Round 1: Washing Out

In addition to observed data, Bayesian inferences require two sorts of assumptions, the likelihood and prior. The assumption of likelihood—or in our terminology, the assumption of a statistical model or "probabilistic kind"—is a common one made in any parametric method, including classical statistics and model selection. The use of prior probabilities, in contrast, is unique to Bayesian statistics and has traditionally been the epicenter of the heated discussion of the pros and cons of Bayesian statistics. Before embarking on this issue, let's consider the following example, which illustrates the importance of priors in Bayesian inference.

> Alice goes to a hospital for a medical diagnosis, and to her dismay she tests positive for a certain disease. The doctor explains that the test kit used for the diagnosis is accurate and can detect 95 out of 100 cases, while it gives a false positive (i.e., diagnoses a healthy person as positive) in only 1 out of 10 cases. Now, what is the probability that Alice is actually ill?
>
> To compute the answer we need a prior probability. From what she has heard so far, Alice estimates the incidence rate of this disease to be about 1%, so that one out of a hundred people has this disease. Letting h denote the proposition that she is actually ill and e the proposition that she tested positive, we have $P(h) = 0.01$, $P(e|h) = 0.95$, $P(e|\neg h) = 0.1$. Then the posterior probability is
>
> $$P(h|e) = \frac{0.95 \times 0.01}{0.95 \times 0.01 + 0.1 \times 0.99} \sim 0.088.$$
>
> Alice is totally shocked by the thought that there is about a one in ten chance that she has the disease.
>
> But is Alice's conclusion really reasonable? Suppose that in reality Alice had greatly overestimated the prevalence of the disease, the true rate being just one out of a thousand. With this correct prior probability $P(h) = 0.001$, the posterior becomes only 0.009. Thus Alice's probability of being ill is less than 1% even if she tested positive.

Concluding a high posterior probability just by looking at a high likelihood as Alice does here is called the *base rate fallacy*. This fallacy vividly illustrates the need for an appropriate prior distribution in justifying the conclusion of a Bayesian inference. Moreover, it casts a shadow of doubt on the objectivity of Bayesian inferences. Let us suppose in Alice's case that the true incidence rate is completely unknown. Then, if two doctors had different takes on the prior distribution, they may end up with different, possibly opposite, diagnoses in the face of the same positive test result. This seemingly implies that Bayesian inferences crucially depend on agents' subjective opinions and do not reach objective conclusions.

The Bayesian's standard answer to this criticism is as follows (Edwards, Lindman, and Savage 1963; Earman 1992). True, we do not reach an objectively

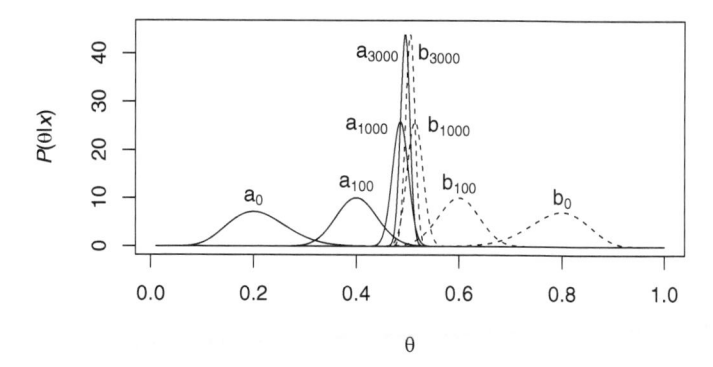

FIGURE 2.2 The "washing out" of priors with data. The solid line a_0 and dashed line b_0 represent, respectively, two different prior distributions on the parameter of a Bernoulli distribution. The a_n, b_n respectively show how these distributions change as we carry out more trials, half of which are successes, for $n = 100, 1000, 3000$. One can see that the two chunks of posterior probabilities approach each other as the data increase, and mostly overlap after the 3000th trial.

justified conclusion in just one shot. Bayesian inference, however, is a process of updating beliefs, and it is by repeating this process that we can arrive at the right conclusion. In our lottery example in Section 2.1.1, we updated the posterior probability of the hypothesis h_A—that the box is A—to 56%, given the evidence of having drawn one blank ticket. We can then draw a second ticket, this time using the posterior we obtained as the new prior. Repeat this process, setting the prior distribution of the nth draw to the $n-1$-th posterior, and after a sufficiently large number n of trials, the Bayesian inference will eventually lead us to the same single posterior distribution, regardless of what priors we began with. Figure 2.2 illustrates this "washing out" of priors using the Bernoulli distribution as an example. The figure shows that the opinions (posterior distributions) of two scientists, a and b, who start with very different prior distributions (the outermost solid line a_0 and dashed line b_0) approach each other as the number of observations is increased, until finally they agree to a large extent. In general, as we obtain more and more data, the effect of priors gets washed out and the posterior distribution converges to the true parameter value (0.5 in this example). Hence, subjective disagreement prior to an inquiry does not pose a serious problem if we have enough data, or so Bayesians argue.

This process of accumulating data can be understood as a sort of justificatory "regress," similar to the one we saw in internalist epistemology. The premise or prior of the nth inference is justified by the conclusion or posterior of the $n-1$-th inference, whose premise in turn is justified by the preceding Bayesian updating, and by continuing this chain of justification, a Bayesian agent aims to justify the whole process of inference. But how long is this chain? Many epistemologists think that it continues indefinitely, and that such an infinite regress cannot justify anything.[7] In contrast, statisticians think that an infinite regress can ultimately justify the conclusion. The basis of this claim is the law of large numbers that we saw in the previous chapter. The law guarantees that unless we assign zero probability to the true hypothesis, our posterior will converge to the true distribution as the number of trials approaches infinity, regardless of what priors we begin with. Some may prefer to view this process of accumulating data as "progress" rather than "regress," but this is just a difference in perspective: the idea that one can reach the truth through an infinite procession of inferences means, when viewed backwards, that the conclusion thus attained is justified by the infinite stock of inferences. At any rate, what is important is that there is a theoretical proof to the effect that the posterior distribution will eventually converge to the truth should one repeat the process of justification indefinitely (Earman 1992, ch. 6). This means that Bayesian justification is asymptotically truth-conducive after all.[8]

Justification of Priors, Round 2: Non-Informative Priors

But all this is valid only if we can afford an infinite or sufficiently large number of trials. In practice, we never have an infinite amount of data; usually we don't even have anything close to that, and in such cases there is a high chance that the prior distribution we took as the premise is not completely washed out but remains to affect the accuracy of the conclusion. In such realistic situations where one can afford only a finite chain of justification, one must choose an appropriate prior distribution as the starting point. The chosen prior is then expected to serve as a base premise, or foundation, for sustaining the subsequent updating process. In philosophy, such a strategy is known as *epistemological foundationalism* (Hasan and Fumerton 2018). Foundationalism assumes that among our beliefs there are "basic" ones that do not require any further justification, but rather serve as the ultimate resource for the justification of all other beliefs. The obvious question, then, is: what are these basic beliefs? If a belief is basic in the aforementioned sense, it must draw its justification solely from itself, without help from any other beliefs. But how is that possible? Two possibilities are conceivable. One is to grant that there exists a form of knowledge that is certain *a priori*, independently of any experience. Some might count beliefs in mathematical propositions such as "1 + 1 = 2" as candidates for such knowledge, for they do seem certain by themselves, without the need for any empirical justification. Alternatively, some philosophers like Descartes thought that our beliefs (or "ideas" in his terminology) about the existence of the self and God have an *a priori* certainty even greater than that of mathematical beliefs, and tried to ground our entire system of knowledge on that firm basis. The second possibility is to adopt an *a posteriori* strategy, and to accept our beliefs in basic sense perceptions like "I now see such-and-such a color" as primitive and certain by themselves.

Likewise, the Bayesian justification of priors may take two directions. The *principle of indifference* is the most standard *a priori* strategy for justifying prior distributions without resorting to experience (e.g., Gillies 2000; Williamson 2010) This principle says that, in the absence of any prior information about the hypotheses under consideration, we should regard them as equally likely and assign to them the same probability.[9] The prior distribution thus obtained is called a *non-informative prior*. The distribution that assigns the probability 0.5 to both boxes A and B in the scenario we considered previously is an example of a non-informative prior. When we have no specific information about what is inside the box, the non-informative prior for the hypotheses about the parameter (i.e., the probability of winning) is the uniform distribution over $0 \leq \theta \leq 1$. When we want to use a normal distribution instead of a uniform distribution as the prior, we can make it a uniform-like flat distribution by setting a very large variance. Although this may not be non-informative in a strict sense, it provides a good enough approximation in practice.

But how can such *a priori* premises warrant the correctness of a conclusion about an empirical hypothesis? This question also arises for any foundationalist with an *a priori* slant. To answer this question, we first need to be clear as to what is intended by the principle of indifference. The gist of the principle, as a guide for setting a prior distribution in the absence of background knowledge, is above all to rule out the biases of individual agents and settle on the same and supposedly most neutral starting point. This intersubjective agreement ensures that any Bayesian updating process using the same dataset will bring us all to the same conclusion or degree of belief. The intersubjectivity thus achieved, however, does not by itself imply that we have reached an objective accordance with the external world—we may have all come to a wrong conclusion together! This means that the principle of indifference does not suffice to establish the truth-conduciveness of Bayesian inferences, unless it is combined with another assumption. The other assumption we need is that the other premise, the likelihood, is correct. Indeed, the guiding idea behind the non-informative prior is to let the conclusions of Bayesian inference depend entirely on empirical evidence rather than subjective and arbitrary opinions. How the data affect posteriors, however, depends on the likelihood function, and which likelihood function we should use—that is, which probabilistic kind aptly captures the inductive problem at hand—is an empirical question *par excellence*. Hence, although the principle of indifference may be a nice start, it cannot be the whole story; to establish the truth-conduciveness of Bayesian justification, it must be coupled with a further empirical assumption.

Moreover, there has been skepticism as to whether the principle of indifference actually fulfills even this restricted goal of establishing intersubjective agreement in an *a priori* fashion. Even if one grants that the non-informative prior is the most neutral and unbiased, one may still ask why we need neutrality to begin with (Williamson 2010, ch. 3). If intersubjectivity is all that matters, why can't we adopt other methods for fixing the prior, say by fiat or voting? Another technical concern is that non-informative priors are not invariant under variable transformations, so that they cease to be non-informative when the object is described in terms of different variables. This problem has been known under the names of the "wine/water paradox" or "Bertrand paradox" and points to the interesting fact that even with the assumption of no information, the priors of Bayesian inferences cannot rule out arbitrariness in variable choice. Readers interested in these issues are referred to Gillies (2000), Williamson (2010), Childers (2013), and Rowbottom (2015), which have detailed treatments of this topic.

Justification of Priors, Round 3: Empirical Bayes

The principle of indifference provides an *a priori* justification of prior probabilities in the absence of any supporting experience. If, on the other hand, something is known about the hypothesis under question, one may well consider integrating this information into the prior distribution. In the previous example of medical diagnosis, we used the background knowledge that the incidence rate is 0.1% as our prior. Calibrating priors with the data at hand instead of using non-informative priors gave a far better inference in this particular case. This method of adjusting the prior distributions *a posteriori* in accordance with background knowledge or known data is a feature of what is called *empirical Bayes* (Efron and Hastie 2016, ch. 6).

Empirical Bayes is motivated by a natural and reasonable postulate, namely, that we ought to adjust our degree of belief in a hypothesis in such a way that it matches the actual chance of the target phenomenon's occurrence. Philosopher David Lewis dubbed this requirement the *Principal Principle* and took it up as the essential (or "principal") assumption in our application of subjective probabilities to inferences about the objective world (Lewis 1980).[10] When we introduced the concept of subjective probability in Section 2.1, we were indifferent about how we set our degrees of belief as long as they satisfy the axioms of probability, so that it would even be OK to assign a high probability to the blatantly absurd blue cheese moon hypothesis. Given that a strict adherence to the probability axioms will prevent one from being Dutch-booked and will help avoid bets that incur a sure loss, it does ensure rationality to some extent. But this is indeed a very small extent, and some may find it far insufficient as a rule for rationality. Indeed, you might be hesitant to call someone very rational if that person attaches a high probability to the belief that "It will snow in Hawaii next New Year's Day," even if his or her probability assignment complies with the probability axioms. This is because such a person will *very likely*, if not surely, lose money if a bet is made based on this hypothesis. If rationality requires us to avoid not only sure losses but also likely risks, we need to adopt an additional constraint to the effect that our degrees of belief match the actual chances of the target phenomena—this is precisely what the principal principle says.

The requirement of the principal principle should sound natural, maybe too natural, so much so that some may wonder whether it deserves such a pompous name: who on the earth wants their degree of belief to be discordant with the actual chance? Justifying this seemingly obvious requirement, however, is not as easy as it first appears. For one, we do not yet have a clear definition of the "actual chance of the occurrence of an event." What is it if not a degree of belief, and what does it mean to match these two notions—the subjective probability and the allegedly objective chance? The first thing to note is that probabilities in the Bayesian semantics refer to degrees of belief, so that the "chance" in question here should be distinguished from the probability we have been talking about. What is it, then? The most obvious candidate would be the actual frequency of the target phenomenon, but defining the "chance" in terms of frequencies is not so easy as it may appear. Let us explain the difficulty with the previous example in which we calibrated the prior probability of a disease using its past incidence record. Suppose that the record contained data of 1000 patients, out of which 10 had the disease. Should we then use all of this data and fix our prior to 1%? That sounds reasonable, unless we independently knew that men and women have different incident rates, in which case we should ignore or at least discount the data from male patients in diagnosing Alice's condition. The problem is that the same worry may arise in principle with any categorization, such as age, place of residence, medical history, or zodiac signs. It is not *a priori* clear which among these categories should be considered, and if we decide to stratify the data with all the categories, we may end up throwing away as irrelevant all cases other than Alice herself. The moral of this story is that the "chance of occurrence" can be defined only with respect to a certain *reference class*, and that it is not necessarily obvious which class is appropriate for a given problem (Hájek 2007; Bradley 2015).

Furthermore, even if we are somehow able to define the actual "chance of occurrence" based on frequencies with respect to a certain reference class or on other grounds, the problem of how it relates to probabilities in Bayesian statistics still remains. The fact that the "chance of occurrence" refers to an objective property of an event—such as its actual frequency—means that it differs in nature from probability in the Bayesian context, which, we recall, was defined as an agent's subjective degree of belief. What the principal principle purports to do, then, is to justify degrees of belief in terms of something outside of beliefs.[11] But how is such a justification possible, if Bayesian justification is by nature an internal relationship among subjective beliefs, as previously discussed? At any rate, a coherent assignment of probability values does not seem to help us here. The proponents of the principal principle, then, need to invoke a concept of justification different from that used within the standard Bayesian framework, or just accept it as an unjustifiable dogma. This indicates a conceptual, if not practical, difficulty underlying empirical Bayes. Calibrating our subjective probabilities and other premises with the data at hand does seem very natural, and arguably it enables us to make better inferences if we can expect to have good background knowledge, as in Alice's case. A theoretical justification of such a practice, however, cannot be found in the internalist framework of Bayesian statistics and must come from outside.

In fact, this is a general problem faced by any internalist epistemology with a foundationalist slant. Foundationalists believe that there are basic beliefs that stop the justificatory regress without itself needing a further justification. But are there really such beliefs that justify themselves, like the unmoved mover? If they exist, the most obvious candidates would be sense perceptions like colors or figures appearing in one's field of vision. Indeed, the visual image I am having now may well appear to justify my belief that I am looking at a computer screen now. But since justification, by nature, is an inferential relationship among propositions, in order to carry out its justificatory job the sense data must have a certain propositional content, like "I'm looking at a black spot now." Once rendered in this way, however, the truth of the proposition comes into question, and I bear the burden of dispelling the doubt as to whether I am really looking at a black spot rather than hallucinating. But that would require new evidence (e.g., that my vision is well-functioning), and hence, the belief in question can no longer be said to be basic. The foundationalists' attempt to stop the regress with basic sensory experience thus fails. The philosopher Wilfrid Sellars called such immediate experience the *given* and dismissed it as a myth (Sellars 1997). That is, there is no such thing, either in the form of beliefs or other perceptual experiences, that can conveniently do the double duty of securing its own empirical correctness and serving to justify other beliefs too.

Exactly the same conceptual problem arises for empirical Bayes. The past data or frequencies used to adjust a prior distribution in the empirical Bayes approach play the same role as the sense data used to justify basic beliefs in foundational internalism. In order for the data (which is literally the Latin word for "given") to serve as a basis for subjective Bayesian reasoning, however, they must themselves take the form of beliefs; that is, it must be the belief that certain data are observed that does the real justificatory work. If so, we need to consider the probability of this belief—we have to assign a certain degree to the belief that such and such data are obtained. However, such an assignment would in turn require justification from other beliefs, especially those about the likelihood and further prior, and thus we are faced with a regress again. Admittedly, such an infinite regress is merely an in-principle possibility which hardly, if ever, arises in actual practices—but that is only because the regress is cut off, usually implicitly, by a fiat like the principal principle.

Just as a justificatory link that is supposed to connect the given data with a basic belief cannot be found in the logic of internalist epistemology (assuming the Sellarsian criticism holds), this principle at the basis of the empirical Bayes approach does not have a justification within Bayesian statistics. Setting up observed frequencies as a basic prior exempt from any further probabilistic justification is more like a declaration or agreement than an empirically verified practice or a mathematically derived protocol. Moreover, if one goes on to use the same data to calculate posterior probabilities, one ends up using the same data twice in a single inferential procedure. This is called *double dipping* and violates the so-called *likelihood principle*, one of the central tenets of Bayesian statistics (Section 3.3.3). Hence, although the principal principle and empirical Bayes appear to provide an intuitive and effective way of ensuring the truth-conduciveness of the Bayesian brief updating process, it is hard to come up with a theoretical justification of these approaches, at least within the internalist Bayesian framework of "relationships among beliefs."

Justification of the Likelihood

The discussions in this section thus far have focused on the justification of prior distributions. Now let us turn to the other principal assumption of Bayesian inferences, the likelihood. The likelihood, or the probability of data under a given hypothesis, is determined by the probabilistic kind that models the stochastic process under consideration. As we saw in Section 1.2.3, identifying a probabilistic kind or distributional family allows us to express the probability of data as a function of parameters of the distribution. The likelihood term in Bayes' theorem expresses this functional relationship in the form of the conditional probability $P(e \mid \theta)$, in which the parameters θ determine the distribution of data e. Clearly the form of this functional relationship should depend on the nature of the inductive problem at hand. The justification of likelihoods, therefore, boils down to the task of choosing the appropriate probabilistic kind that effectively captures the underlying uniformity of nature.

But what does it mean for a model to be appropriate? This is in fact an intricate question to which we will return later in Chapter 4, where we discuss model selection. But here we can take a simple realist stance: a statistical model, or a probabilistic kind in our terminology, is appropriate when it faithfully represents the target phenomenon; in other words, it "carves nature at its joints" as any decent natural kind is expected to do. In the probability-theoretic context, this means that the assumed distributional family includes the true distribution, or one which is fairly close to it, so that a proper tuning of its parameter(s) will make it a fairly faithful picture of the data-generating process. Since inferential statistics carries out inductive reasoning through this underlying uniformity (Figure 1.2), and since in parametric statistics the uniformity reveals itself to us only in the form of parameters (or posterior distributions thereof) of the assumed distributional family, this realist requirement should strike us as a reasonable one.

The next natural question is: how do we know whether the probabilistic kind or likelihood function used in a specific study is appropriate in the aforementioned sense? In the case of priors, we had two routes for justification—an *a priori* one and an *a posteriori* one. Unfortunately, there is no *a priori* route for the likelihood, for no armchair thinking can ensure that our model correctly captures the uniformity of nature. A probabilistic kind is an empirical hypothesis about the data-generating process, which can be evaluated only by actually looking at the process through data. This evaluation procedure is called *model checking* and comes in two varieties (Gelman et al. 2004, ch. 6). The first type of check precedes the application of Bayes' theorem and sees whether the observed data actually follow the assumed distributional family using some statistical tests, like a normality test. The other type is done after the Bayesian inference and examines the match between the posterior predictive distribution (Section 2.2.3) obtained from the inference and the sample distribution (i.e., the actual distribution of data). If these checking methods detect a large discrepancy between the data on the one hand and the assumed distributional family or posterior predictions on the other, we may reject the model's assumptions as false.

It is noteworthy that these checks are not part of the standard Bayesian probability calculus of posteriors from priors, but rather additional procedures to be conducted before or after the analysis. In most cases, checks are carried out by statistical tests (discussed in Chapter 3) or visual inspections of distributional forms or sample statistics. These are epistemic processes distinct from Bayesian inference, understood as a probabilistic derivation of degrees of belief, and as such they do not produce conclusions in the form of probabilities—for example, they do not say anything like "the posterior probability that the assumed model is correct is such and such." Rather, they are a kind of hypothetico-deductive inference that makes a judgment about the hypothesized statistical model by comparing its logical consequences (posterior predictive distributions) with the actual data, and in this sense they are more akin to the procedures of classical test theory to be discussed in the next chapter (Gelman and Shalizi 2012).[12]

The *hypothetico-deductive method* determines the truth or falsity of a hypothesis H that implies a certain prediction E. If the implication is deductive ($H \supset E$), a failed prediction ($\neg E$) falsifies the hypothesis ($\neg H$). But the method does not go that easily when the hypothesis is formed by a conjunction of more than one premise, for in this case there is no principled way to single out the premise(s) responsible for the observed discrepancy. This is the general point made by the famous *Duhem–Quine thesis*. Duhem, and subsequently Quine, argued that since most if not all scientific theories require multiple auxiliary hypotheses in order to produce a concrete prediction, one cannot conclude the falsity of a theory even if one of its predictions turns out to be false. The real culprit may be one of the auxiliary hypotheses, and it is always possible to save the main body of the theory by an *ad hoc* tinkering of these.

W. V. O. Quine, one of the most influential philosophers of the 20th century, extended this idea from scientific theories to knowledge in general and argued for *epistemological holism*, according to which no belief exists in isolation—rather, all beliefs are connected with other beliefs, forming a network (Quine 1951). In this picture, confirmation of knowledge cannot take place in a piecemeal fashion, by picking up a particular belief one by one and checking if it matches with a piece of experience. If all of our beliefs are connected directly or indirectly to each other through logical or empirical laws, as holism holds, then what is tested by "the tribunal of sense experience" (Quine 1951, p. 38) is our entire network of beliefs together. A systematic discrepancy in such a holistic test urges us to revise *some* part of our belief system but does not pinpoint *where*: there is always more than one way to fix the network, and there is no single correct answer.

The same issue arises in model checking. Posterior predictive distributions are derived from a conjunction of numerous premises, including assumptions not just about the prior distribution and likelihood, but also about IID, experimental design, observation processes, data handling and processing, and so on. When we put the predictions derived in this way before "the tribunal of observed data," we are actually testing the whole set of such assumptions, and there is no logical criterion for pinning down the responsible element when a discrepancy arises. The cause of a discrepancy may lie in some of the modeling assumptions (prior distribution or likelihood), or in extra-theoretical procedures (e.g., an inadequate data-gathering process or miscalculation). An analyst has to examine each of these possibilities and take the appropriate measure, which may be a re-specification of the model, revision of the experimental design, or recalculation. The process of model checking is thus more like a trial-and-error process of fumbling for a better model than a straightforward protocol guided by some external standard (Gelman and Shalizi 2012). This is reminiscent of Otto Neurath's famous metaphor, which likens scientific practice to the ship of Theseus. According to Neurath, a scientific theory or model is like a ship sailing in the wilderness of sea. As sailors, scientists must keep fixing the ship that has been damaged by the raging waves of experiments, in order to continue the journey as far as possible. Repeated repairs may transform the ship into an utterly different shape. And no matter how hard their ship gets wrecked by accumulated anomalies, scientists cannot just step out and overhaul the whole ship from the outside. They have no other choice but to keep devising the best patches available on board to continue their journey. This metaphor applies nicely to Gelman et al.'s picture of elaborating a statistical model through repeated posterior model checks. They emphasize that Bayesian inferences do not end with a derivation of a posterior distribution: rather, the critical part of the statistical analysis lies in the subsequent process of repeatedly modifying the obtained model using existing or new datasets. This kind of process of testing and mending necessarily takes on an *ad hoc* flavor. At any rate, one cannot just "step out" of their statistical model and compare it with the ground truth from an external, god-like perspective. That is the fate of Neurathian scientists *qua* eternal sailors, and in this respect, statisticians are no exception.[13]

The resulting perspective on the practice of Bayesian statistics prompts us to rethink the characterization of Bayesianism developed throughout this chapter as a foundationalist epistemology of the internalist kind. On this internalist interpretation, the validity of Bayesian inference hinges on two factors: the sound application of the logical rule (i.e., Bayes' theorem) by which one derives one belief from another, and the adequacy of the premises that serve as the basis of the inference. In line with this scheme, traditional Bayesians have devoted much effort to justifying (a particular form of) prior distributions, just as internalist epistemology has tried to justify our basic beliefs. Such a foundationalist picture, however, does not square well with Gelman et al.'s approach, which highlights the continuous process of model checking as the key element of Bayesian inference. This new picture is much more holistic in spirit, and as such it does not require any "basic belief" that would serve as an Archimedean point for the entire inferential process. Quine (1951) argued that even mathematical objects and rules are not eternal truths engraved on stone, but are mere instruments that we use for understanding the world and predicting the future. Likewise, prior distributions are not fundamental premises but just one of the tools (a regularization device; see Chapter 4) we use to prevent models from overfitting the data (Gelman and Shalizi 2012). According to this holistic interpretation, Bayesian inferences depend not just on prior distributions but also on the likelihood and other theoretical as well as empirical assumptions, and it is the entire network consisting of all such heterogeneous elements that is assessed by model checking. In this sense, those assumptions concerning data-gathering and handling processes, which are often considered "outside" of statistical modeling proper, are on a par with the fundamental mathematics of Bayesian statistics, and call for equal care in the inferential practice.

Such a holistic assessment may go beyond the internal system of beliefs. Recall that "beliefs" in the Bayesian framework, strictly speaking, are limited to elements in the sample space that can be assigned a subjective probability (see Section 2.1). On the other hand, the various assumptions about models, experimental design, measurements, and other factors are usually not part of the sample space, and as such are not things that are assigned probabilities or "degrees of belief." A diehard subjectivist might still argue that such assumptions must exist in the mind of an analyst in the form of belief. But this response does not cut much ice, for even if we grant that these assumptions are indeed "beliefs," they differ in nature from beliefs in the technical sense, defined in the formal framework of the semantics of subjective probability, and as such they do not admit of degrees measured by a probability value. The metaphysical rejoinder, therefore, does not affect the point that Bayesians must take into account those assumptions that cannot be rendered into beliefs in the technical sense. This implies that the process of inductive reasoning as a whole cannot be confined only to the system of beliefs internal to an epistemic agent and the logical relationships between them. Even if the posterior distributions can be derived from a belief calculus using Bayes' theorem, checking the empirical adequacy of the calculated result calls for a reference to external assumptions. Holistic Bayesians on board Neurath's ship will have their eyes open to the external world beyond their beliefs.

2.3.4 Summary: Epistemological Implications of Bayesian Statistics

A statistical method is an epistemological procedure for justifying scientific hypotheses on the basis of data. Building on this view, in this chapter we analyzed Bayesian statistics as an internalist epistemology that aims to justify our degree of belief in a given scientific hypothesis in terms of the data, likelihood, and prior probabilities. This view of justification as an inferential relationship is motivated by the principal tenet of internalism, namely, that only those beliefs appropriately inferred from legitimate evidence count as knowledge. Although internalism does seem to capture an important aspect of our conception of justification, it faces the difficult problem as to how and why justification, understood in this way as a subjective relationship, can ever be truth-conducive—that is, on what grounds can one say that internally justified beliefs are also *objectively* true? Exactly the same kind of problem has been raised against traditional Bayesianism. Karl Popper, a philosopher of science known for his celebrated falsificationism, is one of those critics. He denounced Bayesian statistics as mere psychologism, preoccupied with calculations of the subjective opinions of individual scientists, which is utterly inadequate for a scientific investigation about the objective structure of the world (Popper 1959). Popper's criticism is targeted precisely at the truth-conduciveness of Bayesian justification. Bayes' theorem may allow epistemic agents to adjust their beliefs in a coherent fashion; but whether these beliefs are objectively correct or not is a different question, one that is arguably much more important in scientific contexts.

We have seen two responses to this criticism from the Bayesian camp. One is to fully admit the subjective nature of Bayesian justification, but to argue that repeated application of the justificatory process through numerous trials or observations will bring the posterior distribution closer to the truth. The sequential process where one justifies a new prior using the posterior distribution of the previous trial may be thought of as a kind of inferential "regress." While most philosophers have been concerned that such a regress will not bring us anywhere, the law of large numbers guarantees that the infinite "regress" of Bayesian justification will converge to the truth, and this gives the Bayesian a rationale for this type of response.

However, iterating the justificatory procedure *ad infinitum* or even sufficiently many times is not a realistic option in many situations. Practical limitations in data, time, and resources in real-life problems call for an alternative strategy, which is to justify the prior distribution and likelihood—the premises of Bayesian inference—in some way or another. This strategy either invokes extra postulates, like the principle of indifference or Lewis' "principal principle," in order to stop the justificatory regress, or checks the adequacy of the premises by comparing the result of the Bayesian estimation with the data in a hypothetico-deductive way. The latter two strategies are *a posteriori* in nature in that they evaluate the adequacy of the assumptions (prior distributions and likelihoods) in light of observed data. The "justification" of the premises obtained in this way, however, differs in nature from the concept of justification proper to Bayesianism, where it is understood as a mathematical/deductive relationship among propositions, and this leaves open the question as to in what sense the premises can be said to be justified, if at all.

Here we encounter the fundamental difficulty inherent to internalist epistemology. Internalism, by definition, locates the justificatory process within the mental resources and activities accessible to an epistemic agent. Logic and probability are powerful vehicles for generating inferential relationships among the beliefs held by such an agent. As one might easily expect, however, this strategy faces the difficult problem of relating internal beliefs to external facts. This was exactly the point of the Sellarsian criticism of the "Myth of the Given," which suggests that the process of justifying beliefs can never be completed within a subject. Likewise, if Bayesian statistics is to be confined entirely to the internalist process of justification, this can be done only by taking its premises about prior distributions, likelihoods, and data as "given," not admitting of any further justification. But in this case, one loses all means of answering the skepticism as to whether the resulting estimation or conclusion correctly captures the way the world actually is. In contrast, the idea of model checking introduced in the previous section takes Bayesian inference not as a one-way deduction from the given, but rather as a sort of continuous dialogue between the analyst and nature, in which the predictions of a model are compared with new data to test the model's assumption *ex-post facto*. Since all the assumptions making up the analysis, including not only the priors and likelihoods but also those concerning experimental design and measurement, must be taken into account in such a test, it takes on a holistic character. The goal of statistical analysis, then, is to amend and improve the holistic network of assumptions and beliefs so that it better accommodates experience. At the same time, this forces us to step out of the well-defined mathematical realm of "beliefs" defined on a sample space, and into the unformalized, qualitative problems of grasping the outer world and assessing our experimental and measurement setups. How do these heterogeneous assumptions work together in inductive reasoning, and what role do they play in Bayesian inferences? These are philosophical as well as practical questions, under which lies the time-honored philosophical conundrum of how subjective models can ever latch onto an objective reality.

Further Reading

For philosophical textbooks on the interpretative/semantic issue of probability, see Gillies (2000), Childers (2013), and Rowbottom (2015), which also cover various other interpretations, such as the logical and propensity interpretations, not discussed in this book. It is impossible to single out a textbook on Bayesian statistics: Hoff (2009) provides a balanced overview of the foundations as well as the practical applications of Bayesian statistics, while Gelman et al. (2004) is the standard reference in the field. Bayesian epistemology is often called *formal epistemology*; Jeffrey (2004) and Bradley (2015) give good introductions to this topic. For more in-depth philosophical analyses of Bayesian statistics, Earman (1992) and Howson and Urbach (2006) are classic monographs, and Sober (2008) also devotes a chapter to Bayesianism. McGrayne (2011) gives an enjoyable account of the history of Bayesianism. There are many introductory textbooks on philosophical epistemology, among which Nagel (2014) and Pritchard (2014) are the most accessible.

Notes

1. A contradiction is a proposition that is never true, such as "Today is Monday and today is not Monday."
2. Here we limit our discussion to at most a finite number of propositions. For the case of a countably infinite number of propositions, see Gillies (2000, sec. 4.3).
3. In contrast, the *logical interpretation* of probability asserts that there must be a single correct probability assignment. This view takes probability as denoting a logical relationship, where one proposition is *partially* entailed by another. See Gillies (2000) for details.
4. To be precise, since we are here considering a probability distribution over uncountably many parameter values, what we are really calculating are posterior probability densities. See Section 1.2.2.
5. While our derivation of the posterior predictive distribution here is based on an intuitive argument, a closer look reveals that this derivation too relies on the assumptions of uniformity and a statistical model. Applying the law of total probability to a conditional probability, the posterior predictive probability can be expanded as:

$$P(\tilde{e} \mid e) = \int P(\tilde{e} \mid \theta, e) P(\theta \mid e) d\theta.$$

From the assumption in Figure 1.2, observed data e and unobserved data $\tilde{e}$ are mediated only via the underlying uniformity. Hence, if this uniformity is adequately captured by a statistical model, the two kinds of data can be made independent by fixing the model's parameter—or in a philosophical parlance, the unobserved data $\tilde{e}$ is screened-off from the observed data e by parameter θ. The equality $P(\tilde{e} \mid \theta) = P(\tilde{e} \mid \theta, e)$ thus holds for all θ, yielding the equation in the main text. Here again, a crucial role is played by the assumption of uniformity (the IID condition) as well as the assumption that this uniformity is well-captured by a statistical model.

6. This difference might seem negligible at first sight. One may, for example, propose to set a certain threshold and accept any hypothesis whose posterior probability exceeds that threshold. But this strategy does not work, as is shown by the well-known *lottery paradox*. Imagine 100 lottery tickets, among which only one is a winning ticket. For each ticket, the probability of it being blank is 0.99. Hence, if we decide to accept as a truth any proposition having probability 0.99 or greater, then we end up judging, for every ticket, that it is blank. This, however, contradicts our supposition that one out of the 100 tickets is *not* blank. Note that the same contradiction can be obtained for an arbitrarily stricter threshold by increasing the number of lots. This paradox implies that a set of beliefs that is taken as (a candidate for) "knowledge" according to a certain threshold is not logically closed. This goes against our expectation that knowledge should be logically closed (so that the logical consequence of a piece of knowledge should also count as knowledge), and is thus considered paradoxical.

7. But some epistemologists, called infinitists, do allow justification by infinite regress (Klein 1999).

8. However, whether limit theorems really bear out the objectivity of Bayesian inference even at the limit has been questioned by some (e.g., Glymour 1981). I thank Jimmy Aames for bringing this point to my attention.

9. Or, in its modern formulation, it tells us to adopt the prior distribution that maximizes entropy, which is considered a measure of uncertainty (Jaynes 1957). This is called the *Maximum Entropy Principle*, and the branch of Bayesianism that emphasizes this principle is called *objective Bayes* (Williamson 2010).

10. In the epistemological literature, the degree of belief, called credence, is distinguished from the objective chance of a phenomenon's occurrence. Let us write $ch_A(H) = x$ to denote that some agent A believes that the chance that a phenomenon H will occur is x, where $0 \leq x \leq 1$. The principal principle then asserts that A's degree of belief P_A must satisfy

$$P_A\left(H \mid ch_A(H) = x\right) = x.$$

11. To be precise, the principle attempts to justify the degree of belief of an event in terms of another *belief* that the agent has about the chance of the target phenomenon. But this only pushes the problem back, since one can still ask how such a belief about chance is justified.

12. In many cases, however, model checks are carried out by a simple inspection of the match between the model's conclusion and data. And even if a statistical test gives a poor fit (such as a low *p*-value), this does not logically falsify the model's assumptions, for these are by nature stochastic hypotheses (see Section 3.2.1 of this book and Sober 2008).

13. Gillies (2009) and Sober (2015) also discuss, in different ways, the implications of Neurath's idea for Bayesian inference.

3
CLASSICAL STATISTICS

In this chapter we turn to classical statistics, in particular its theory of testing. Although the idea of statistical testing can be traced back to the early 18th century, when Scottish physician John Arbuthnot set out to test whether the birth rates of boys and girls are equal using London birth records spanning over 82 years (Hacking 2016), its theoretical basis was furnished only in the 20th century, largely by the hands of Sir Ronald Fisher, "the founding father of modern statistics," and later Jerzy Neyman and Egon Pearson (Karl Pearson's son). Their seminal work pushed classical statistics into the mainstream of inferential statistics, where it virtually reigned as *the* standard statistical method until its hegemony was challenged toward the end of the 20th century, as Bayesian statistics came to gain popularity thanks to the growth of the power of computers. Even today, some of the characteristic terminology of classical statistics like "statistical significance" and "*p*-value" belong to common vocabulary and appear not only in scientific literature but also in the popular media. Despite this popularity, however, the exact meaning of these terms, as well as the underlying ideas of classical statistics, is much less understood. Indeed, the logic underlying statistical testing is rather intricate and less intuitive than the Bayesian concept of updating beliefs based on data, and this sets a high bar for understanding classical statistics.

Classical and Bayesian statistics, both being part of inferential statistics, share the same goal of capturing the probability model that underlies the observed data. They disagree, however, in the following two respects. First, their inferential practices are rooted in different conceptions of probability. While Bayesians interpret probability as a subjective measure of the degree of belief, in classical statistics it is defined objectively, as the relative frequency of a sequence of events. Second, they have different takes on what constitutes inductive

DOI: 10.4324/9781003319061-4

reasoning. In Bayesian statistics, inductive reasoning means the adjustment of beliefs about a probability model on the basis of data. Classical statistics, on the other hand, makes an inference by framing some definite hypotheses about the probability model and then testing them vis-à-vis the data. This implies that Bayesian and classical statistics embody distinct epistemologies, which differ not just in their methodology but also in their very conception of inference, giving different answers to the question of what inductive reasoning really is. With this in mind, we begin this chapter with a brief review of the semantics of probability in classical statistics. We will then look at its methods and epistemological implications.

3.1 Frequentist Semantics

Probably the most familiar conception of probability is that it refers to the frequency with which a type of event occurs. To say that a coin has a one-half chance of landing heads means, according to this common idea, that when we toss the coin repeatedly, the ratio of heads and tails will be equal. *Frequentism* takes this to be the very meaning of probability and argues that the probability is nothing but a relative frequency, i.e., the number of occurrences of a given type of event in repeated trials, divided by the total number of trials. But this idea does not work as it stands, for the number of trials we can conduct is always finite, whereas finite sequences of the same kind of trial rarely result in the same frequency. Toss a fair coin 100 times and record the number of heads, and your count will rarely be exactly 50; in the next set of 100 tosses you will likely get a different number. This apparently leads us to the troublesome conclusion that the probability of a given type of event—here, tossing the same coin—cannot be uniquely determined. This can be avoided by extending the sequence of trials from a finite one to an infinite one. The relative frequency of a given type of event may fluctuate in finite sequences, but it should converge to a single value if it is repeated indefinitely. The probability of an event-type is defined as the ideal limit point of such an infinite sequence. This convergence process is demonstrated in Figure 3.1, which is a record of my tossing a coin thousands of times— well, not really; it's actually a simulation run on my computer. The plot shows that the relative frequency of heads approaches one-half as the trial continues, and from this we can conclude that the probability of landing heads when we toss a fair coin (or to be exact, its simulation) is one-half.

Let's exploit this idea using the notion of a probability model that we defined in Chapter 1. Recall that our sample space Ω contains all the possible outcomes of a trial, say a coin toss, where an event like "landing heads" is expressed with a random variable defined over Ω. Let us denote a single coin toss as $\omega \in \Omega$. What we are interested in is the sequence of coin tosses $C_n = \{\omega_1, \omega_2, \ldots, \omega_n\}$, which is called a *collective*. If we express the event

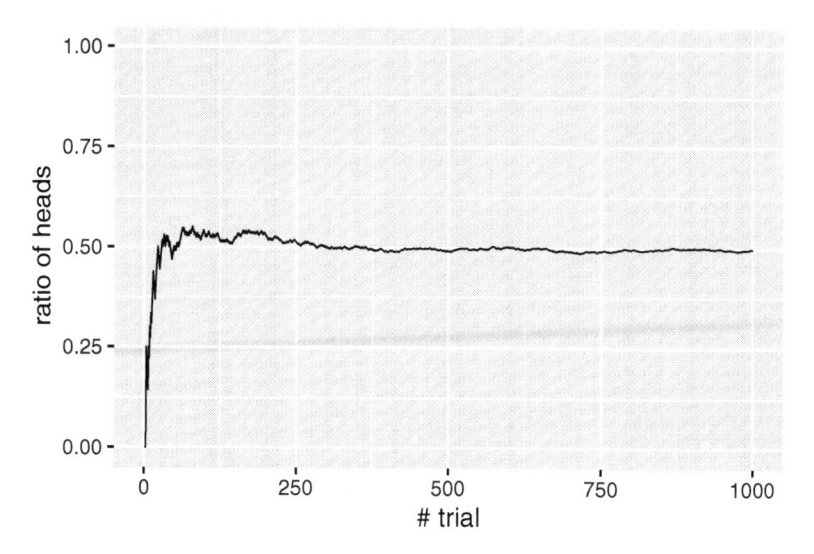

FIGURE 3.1 Coin toss simulation, which shows that the relative frequency of heads approaches to one-half as the number of trials increases.

of getting heads in the ith trial using the random variable $H(\omega_i) = 1$, then $H = 1$ is the set of all tosses in which we land heads.[1] Then the probability of landing heads is defined as the limit as we increase the number of trials n indefinitely, i.e.,

$$P(H = 1) = \lim_{n \to \infty} \frac{|H = 1 \cap C_n|}{|C_n|}$$

where $|A|$ is the cardinality (roughly, the number of elements) of the set A. That is, the desired probability is the number of heads in the sequence C_n divided by the total number of trials, provided that the trial goes on indefinitely. It is easy to check that the probability defined in this way satisfies the probability axioms. The relative frequency of an event A will never be smaller than 0 or larger than 1, so that $0 \leq P(A) \leq 1$, satisfying the first axiom. If we take the whole sample space Ω as an event, the numerator on the right-hand side becomes $|\Omega \cap C_n| = |C_n|$, in which case $P(\Omega) = |C_n|/|C_n| = 1$ in the limit $n \to \infty$; hence the second axiom is satisfied. Finally, if A and B are mutually exclusive, we have $|A \cap C_n|/|C_n| + |B \cap C_n|/|C_n| = |(A \cup B) \cap C_n|/|C_n|$. Taking the limit, we can confirm a simple case of the third axiom, $P(A \cup B) = P(A) + P(B)$.[2]

The biggest advantage of frequentism is that it defines probability objectively in terms of frequencies, which are directly observable. Recall that probabilities in the subjective interpretation are based on the subjective beliefs of individuals. What probability values an epistemic agent assigns to their beliefs is totally up to that agent as long as those values satisfy the probability axioms, and so there is no guarantee that these values are consistent among different individuals. If, on the other hand, probability is defined as a frequency which in principle can be observed by anyone attending to the trial, its value should be uniquely determined according to the way the world actually is. Frequentists take this objectivity and communal character as a big benefit of their interpretation. There is a catch, however. For one thing, frequentist probability is not an actual relative frequency but an ideal limit which would be observed in an infinite sequence of trials. Given that no real person can complete an infinite number of trials, the final result of these trials remains hypothetical and can be expressed only in terms of the value that *would* be obtained *if* we were to continue the trial indefinitely. This raises the difficult problem as to how one can make a judgment about the result of an infinite number of trials based on a finite number of outcomes. Earlier, I claimed that the trend in Figure 3.1 will converge to one-half as the trial is repeated—but how can I be so sure? The mere fact that the relative frequency observed so far is approaching one-half does not afford any basis for my conjecture. It is possible that the coin suddenly begins landing only tails from the 1001st trial on, with the relative frequency converging to zero. It doesn't matter how long we continue the trials: the turning point of the trend may be the thousandth trial, millionth, billionth, or even later. To deal with this issue, Richard von Mises, who put the frequentist idea on a rigorous theoretical ground, required that the "collective" on which probabilities are defined must be random. Certainly, a sequence that suddenly changes its trend at the nth trial is hardly random. But what is randomness, exactly? This is indeed a deep and fascinating problem, but instead of delving into it here, we simply refer the interested reader to the relevant literature (Gillies 2000; Childers 2013).

An important feature of frequentism is that probabilities can be assigned only to "collectives." This has several implications. The first is that events that cannot be observed repeatedly, like those events that happened or will happen only once in the earth's history, cannot have a probability in the frequentist sense.[3] This marks a contrast to the subjective interpretation, which has no problem talking about the probability of—or one's degree of belief in—events like the extinction of the dinosaur due to the meteorite impact 65 million years ago, or that Kyoto will be sunny on New Year's Day 2050—for subjectivists can just make a bet about these events. Such talk about probability, however, is nonsense according to frequentism, for in order to think about the frequencies of these events, one needs to be able to repeat the entire history of the earth infinitely many times and take records, which hardly makes objective sense.

Second, even for events which can be repeated (infinitely) many times, we cannot meaningfully ask what the probability of each trial or instance of those events is. Although it makes sense to say that the probability of this coin landing on heads is one-half, saying that the probability of this coin landing heads *on the next toss* is one-half violates the frequentist grammar. This is because the particular trial of tossing the coin now is a one-time phenomenon and does not constitute a frequency. It is only when that particular trial is considered as an element of a "collective" of an infinite sequence of coin tosses that we can think about its frequency. Even in that case, the frequency is defined as a property of the set of trials understood as a collective, and not of any of its individual elements. It is not the case that each coin toss has its own "frequency" which accumulates to define the relative frequency of the whole sequence. Or, to rephrase in philosophical terms, the frequentist probability is defined exclusively with respect to a type, and it is a categorical mistake to apply it to concrete tokens.[4]

The last, and probably most important, implication for the practice of statistical inference is that the frequent interpretation forbids any talk about the "probability of a hypothesis" as meaningless. Let us illustrate this by taking the general theory of relativity as a scientific hypothesis. Can we meaningfully think about the probability that this hypothesis is true? If we are to make sense of this probability in the frequentist framework, we would need to imagine a "collective" of universes and ponder how many of them obey Einstein's laws. This sounds like a futile speculation at best, and in any case such a "collective" would never be observable. What we can observe is at most just one sample, this world; and whether the theory of relativity holds or not in this world is already determined, even if it is unknown to us. So there is no room for frequencies here. Likewise for any other scientific hypothesis. "Smoking causes cancer," "This urn contains equal numbers of red and white balls"—these hypotheses state the ways (a part of) the world actually is, and whether known to us or not, they are already fixed on the side of the world (provided we ascribe to determinism and ignore quantum uncertainties). Scientific hypotheses are descriptions of the objective world, which exists as something fully determined. Thinking about the probability or frequency of hypotheses, therefore, is nonsense from the frequentist perspective.

This means that in classical statistics, which stands on the frequentist interpretation of probability, it does not make sense to update the probability or degree of belief of a hypothesis about a probability model on the basis of data, as Bayesians do. The frequentist semantics of probability therefore calls for a different kind of epistemology. This alternative epistemology is embodied in statistical hypothesis testing. Statistical theories of testing make a certain hypothesis about the world and see whether it accords with the data. The hypothesis is rejected or retained based on the goodness of fit. Such a judgment is necessarily fallible, but under certain presuppositions one can calculate the long-term error rate of a particular test, i.e., the frequency of incorrect as well as correct verdicts. Classical statistics looks for the test that minimizes these error rates and makes judgments about the probability model based on this test's results. Let us now look into this epistemological side of classical statistics in the next section.

3.2 Theories of Testing

3.2.1 Falsification of Stochastic Hypotheses

The underlying idea of statistical testing shares some similarities with Popper's falsificationism. As seen earlier, Popper dismissed Bayesian inductive reasoning *qua* belief updating as a naive psychologism which falls short of an objective scientific methodology. His counterproposal is the famous idea of *falsificationism*, according to which science proceeds through a cycle starting with the formation of a hypothesis, followed by testing and refutation, which then leads to the formation of a new, better hypothesis. Scientists first propose a certain hypothesis about the phenomenon of interest. The hypothesis implies several predictions, which are then compared with observed data. If the predictions do not accord well with the data, scientists reject the hypothesis as false and start looking for a better hypothesis. But what if the predictions are successful? Even in that case, one cannot be sure that the hypothesis is right, for concluding so on the basis of a successful prediction amounts to the fallacy of affirming the consequent. The only thing we can say for sure is that the hypothesis survived this particular test, and this does not give us any guarantee for the next test. What scientists should do, then, is derive further predictions from the hypothesis and put it to yet another test by comparing them with new data. Again, passing this new test does not prove the truth of the hypothesis. Indeed, the hypothesis forever remains a tentative placeholder, which may turn out to be false in the future. There is no guarantee that it will reach the truth, and even if it does, there is no way of telling. But at least we can rule out false hypotheses by continuing this survival game. Popper thus described science not as an asymptotic approach toward the truth but as a process of systematically plowing out falsehood.

We noted earlier that it is a logical fallacy to affirm a hypothesis from the success of its predictions. On the other hand, the falsificationist procedure of rejecting a hypothesis on the basis of the failure of its predictions is a logically valid inference called *modus tollens*. In effect, it is the underlying logic of the hypothetico-deductive method, in which one infers the falsity of a hypothesis ($\neg H$) in case the hypothesis implies a certain prediction ($H \supset E$) that does not hold ($\neg E$). But this applies only when the hypothesis logically implies the prediction, in which case the predicted phenomenon must necessarily occur should the hypothesis hold. Only in such a case does a failed prediction logically falsify the hypothesis. Actual scientific hypotheses, however, hardly make such decisive predictions. The hypothesis that smoking causes cancer, for example, does not claim that all smokers inevitably get cancer without exception; it just makes a stochastic prediction to the effect that, *ceteris paribus*, smokers tend to develop cancer more often than nonsmokers. And once we accept that the prediction of a hypothesis is stochastic, *modus tollens* is no longer applicable. Many heavy smokers have lived to a ripe old age (Winston Churchill being one), but that does not refute the cancer hypothesis. In general, one cannot reject a hypothesis even if we observe a phenomenon that is deemed very unlikely under that hypothesis.[5]

To verify this, let's return to the lottery box we used in the previous chapter. Suppose you are surmising that today's box is B, which contains 3 winning tickets out of 10. Suppose further that all three people in front of you draw a winning ticket. The probability of drawing three consecutive wins from box B is $(3/10)^3 = 0.027$; that is, it is a fairly rare event that happens no more than three times out of a hundred. Should you then reject your box B hypothesis, given that something very unlikely according to that hypothesis happened? Of course not. For, by assumption, the box is either A or B, and if the box were A, the winning probability would be even lower (10%); hence, the observation of three consecutive wins, far from rejecting the box B hypothesis, supports it.

We ought to draw two morals from this hypothetical story. One is that the likelihood of any stochastic hypothesis, i.e., the probability that particular data are observed under that hypothesis, can be arbitrarily small. If trials are independent, the probability of the obtained sequence of data is the product of the probability of each trial, none of which exceeds one. Hence, by multiplying them over and over, the probability of the whole sequence becomes smaller and smaller as the sample size increases. Thus, the small likelihood of a hypothesis by itself does not amount to either a falsification or confirmation of that hypothesis. Second, in order to judge the truth or falsity of a hypothesis on the basis of its likelihood, one also needs to consider what would happen if that hypothesis were false. If we take the three consecutive wins as evidence for the box B hypothesis, that is because we assume that if the box were not B, it should be A, in which case the winning probability would be even lower. If the alternative possibility had been yet another box C, 50% of whose tickets are winning tickets, the same result would have given us reason to reject the B hypothesis. In testing a stochastic hypothesis, therefore, we need to consider not just its likelihood, but also the probability of obtaining the same result if that hypothesis were false, that is, the likelihood of its rival hypotheses (Hacking 2016; Sober 2008).

3.2.2 The Logic of Statistical Testing

Based on the aforementioned idea, the Neyman–Pearson theory of statistical testing—which forms part of the theoretical core of classical statistics—conducts a test on a stochastic hypothesis by pairing it with its rival hypothesis. The target hypothesis that we want to test is called the *null hypothesis* (or "null" for short) and denoted by H_0, whereas its rival is called the *alternative hypothesis* (likewise, "alternative") and denoted by H_1.[6] The goal of a test is to decide whether to reject or not reject (i.e., retain) the null hypothesis in the face of observed data. The most standard way to implement this decision procedure is to compare the likelihood of the two hypotheses.

Let's illustrate this with an example. A coin issued in country X has a rounded edge and does not stand vertically. Though imperceptible by unaided eyes, the rounding finish of the coin is not even, so it lands heads only once in four times when tossed. The earliest batch of these coins that were minted, however, were pressed with the front and back on opposite sides due to a defect of the minting machine, and as a result they land heads three in four times. This error coin is very rare and has been traded at a high price among collectors. One day, you go to a flea market and find a fishy guy who trades this coin. His price is very reasonable should it be a genuine error coin. Unsure about its authenticity, you ask the guy whether you can test it by tossing it a few times. He allows you to toss it ten times, but not any further for fear of damaging the coin. You thus set out to conduct a hypothesis test to decide whether to buy the coin or not. Your null hypothesis—the hypothesis you are trying to reject—is that this coin is fake (not a genuine error coin). The more heads you get, the more reason you have for rejecting the null hypothesis. The question, then, is: how many heads out of 10 tosses would you need, at least, to decide to reject the null hypothesis and buy the coin?

The desired threshold is called the *critical region*. If the obtained data fall within this prespecified region, the test rejects the null hypothesis. Deciding how to conduct a statistical test ultimately boils down to setting its critical region. The first thing to be noted, however, is that no matter how the critical region is specified, no test can avoid the possibility of error, as long as the target hypothesis is stochastic. There are two types of possible error:

Type I error: falsely rejecting the null when it is actually true (false positive).
Type II error: falsely retaining the null when it is actually false (false negative).

In the present context, the type I error amounts to being tricked into buying a fake, whereas the type II error amounts to missing the opportunity to get a genuine error coin at a bargain price. The probability that a given test commits the type I error is usually denoted by α, and the type II error by β. Since these error rates depend on how the critical region is specified, the task of statistical test theory boils down to finding the critical region that minimizes these error rates.

3.2.3 Constructing a Test

Let us look into this procedure by actually building a test in line with our example. As we saw in Chapter 1, the number of heads in 10 coin tosses follows the binomial distribution, with the parameter θ representing the probability of the coin's landing heads. The null hypothesis that the coin is fake is H_0: $\theta = 0.25$, while the alternative that it is a genuine error coin is H_1: $\theta = 0.75$. Under this setup, let us first calculate the type I error rate α. Since this is the probability of rejecting the null when it is true, we assume H_0 to be true and seek the probability of getting x heads under this assumption, that is, the likelihood of the null hypothesis for each $X = x$. Let us express this likelihood by $P(X = x;\ H_0)$.[7] From the binomial distribution with $\theta = 0.25$, $n = 10$, we have

$$P(X = x; H_0) = {}_{10}C_x(0.25)^x(0.75)^{10-x}.$$

The upper histogram in Figure 3.2 shows the probability values obtained by substituting the values 0 to 10 for X in the aforementioned formula. Now, we were trying to devise a test that will reject H_0 if the number of heads exceeds a certain threshold x'. Since the upper histogram of Figure 3.2 shows the probability of each number of heads when H_0 is true, the probability that the test erroneously rejects the null—that is, its type I error rate α—should be obtained by summing all the probability values to the right of a certain threshold x'. With this in mind, consider the following three tests:

A. Let us agree to reject H_0 only when $x' = 10$—that is, when all the tosses result in heads. Since $P(X \geq 10; H_0) = 0.00000095$, this test commits the type I error once in a million times at most.
B. Lower the bar and consider a test that rejects H_0 if more than five heads are observed. Then its type I error rate is $P(X \geq 6; H_0) \sim 0.020$, i.e., 2%.
C. Consider an even looser test, which rejects H_0 if more than four heads are observed. Its type I error rate increases to $P(X \geq 5; H_0) \sim 0.078$, i.e., about 8%.

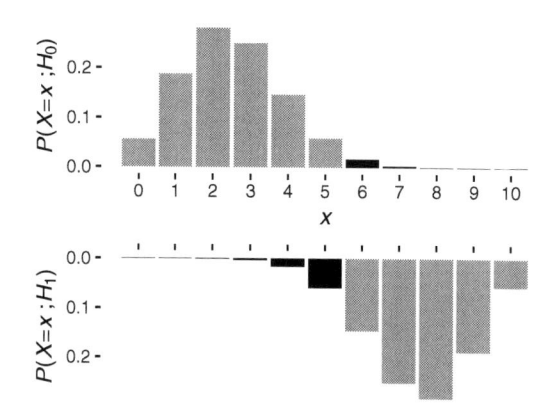

FIGURE 3.2 An example of a statistical test with a simple null hypothesis H_0: $\theta = 0.25$ versus a simple alternative hypothesis H_1: $\theta = 0.75$. The histogram above (below) shows the probability that the number of heads exceeds x, given that H_0 (H_1) is true. The dark gray area represents a test (test B in the main text) which rejects H_0 if more than five heads are observed (the upper part is the type I error rate α and the lower part is the type II error rate β).

The aforementioned comparison makes it clear that the probability of the type I error (of falsely rejecting the null hypothesis when it is true) is determined by how we set the critical region, which is tantamount to selecting a particular test. The type I error rate of a given test is called the *significance level*. A test with a lower significance level (i.e., a smaller α) has a smaller risk of rejecting a true H_0 by chance, which gives us good reason for taking the verdict of rejection from such a test seriously—that is, as significant.

But we should not jump to the conclusion that the test with the lowest significance level, in our case test A, is always the best. For paring down the false positive rate tends to inflate the false negative rate, i.e., the probability of the type II error of failing to reject the false null hypothesis. So let us now calculate this probability. Since a type II error means overlooking the true alternative hypothesis H_1, we assume H_1: $\theta = 0.75$ is true. We are then picturing a probability distribution of the coin toss outcomes that differs from the one we considered earlier. Let us denote the probability of getting x heads under this alternative hypothesis by $P(X = x; H_1)$. Since this is the binomial distribution with $\theta = 0.75$, $n = 10$, we have

$$P(X = x; H_1) = {}_{10}C_x (0.75)^x (0.25)^{10-x},$$

as shown by the bottom histogram in Figure 3.2.

We arranged our test so that it rejects the null when the heads count exceeds x'. In other words, the test does not reject the null when the count is equal to or less than x'. Since the bottom part of Figure 3.2 is the distribution when the null should be rejected, the probability of our test's committing a type II error β can be obtained by summing all the probability values to the left of the threshold x'. Doing the math with the three tests above, we obtain:

A. The test with the threshold $x' = 10$, which rejects H_0 only when all tosses result in heads, has a type II error rate β of $P(X < 10; H_1) \sim 0.944$, which is about 95 out of 100 times.
B. The test that rejects H_0 if more than 5 heads are observed: $\beta = P(X < 6; H_1)$ ~ 0.078, i.e., 8 out of 100 times.
C. The test that rejects H_0 if more than 4 heads are observed: $\beta = P(X < 5; H_1)$ ~ 0.020, i.e., 2 out of 100 times.

What we see here is a trade-off between the type I and II error rates. The test that sets the threshold as high as 10 heads may certainly minimize the false positive rate, but it overlooks most of the cases in which the null should be rejected. On the other hand, lowering the bar to 4 heads may help us detect a false null hypothesis but also increases the risk of erroneously rejecting the null when it is actually true. The inevitability of this trade-off is also apparent in Figure 3.2: sliding the critical region to the right increases the type II error below, while sliding it to the left increases the type I error above.

If the significance level α is the type I error probability of (erroneously) rejecting a true H_0, then $1 - \alpha$ is the probability of (correctly) retaining the true null. Since a test with a high $1 - \alpha$ has a low risk of rejecting a true H_0, we can be confident when it *does* issue a rejection. For this reason, $1 - \alpha$ is called the *confidence coefficient*, or simply *confidence*, of a test. On the other hand, since β is the type II error probability of failing to reject H_0 when H_1 is true, $1 - \beta$ is the probability of not overlooking a true H_1, which is called the *power* of a test. In many cases, we set the alternative hypothesis to the hypothesis we are interested in (in our example, this is the hypothesis that the coin is a genuine error coin). Thus, the power of a statistical test measures how well the test can detect the result we are interested in.

Let's put together the pieces outlined to this point to see how the actual testing practice proceeds. First, we determine the significance level. Generally it is set to 5% or lower, which means that we agree to tolerate mistakenly rejecting a true null up to once out of 20 times. Since test C, which has a significance level of 8%, does not satisfy this criterion, we adopt test B, which will reject the null hypothesis of the coin being fake if more than 5 heads are observed. Now suppose you (finally!) toss the coin and get 7 heads. Since this falls within the critical region of test B, the null is rejected. Or, we can think of it this way. It is somewhat unlikely that we get 7 heads if the null hypothesis, that the coin is biased toward tails, was true. Indeed, calculating from Figure 3.2, the probability of getting 7 or more heads is about 0.35%. The probability of obtaining, under the null hypothesis, a result at least as extreme as the one actually observed is called the *p*-value. Intuitively speaking, the *p*-value measures the "unlikeliness" of data provided the null hypothesis is true. If the result is so "unlikely" in this sense that it is below the level attributable to chance (i.e., the significance level), we reject the null. In this case, we rejected the null using test B, which has a significance level of 2%; but the fact that the *p*-value is 0.35% means that we could also have rejected the null even if we had set the significance level as low as that. In this way, the *p*-value contains information about not only whether or not the null should be rejected, but also the significance level at which it would have been rejected/retained. For this reason, many scientific practices attach high importance to this *p*-value. In this case, one first calculates the *p*-value from the observed data and then makes a decision on the null hypothesis by comparing it with the prespecified significance level.

3.2.4 Sample Size

In the previous example, you were allowed to toss the coin only 10 times. How would the inference improve if you were allowed to toss it 20 times? In this case, the likelihoods of the null and alternative hypotheses are given by binomial distributions with $n = 20$ (Figure 3.3). Since $P(X \geq 8; H_0) \sim 0.12$ and $P(X \geq$

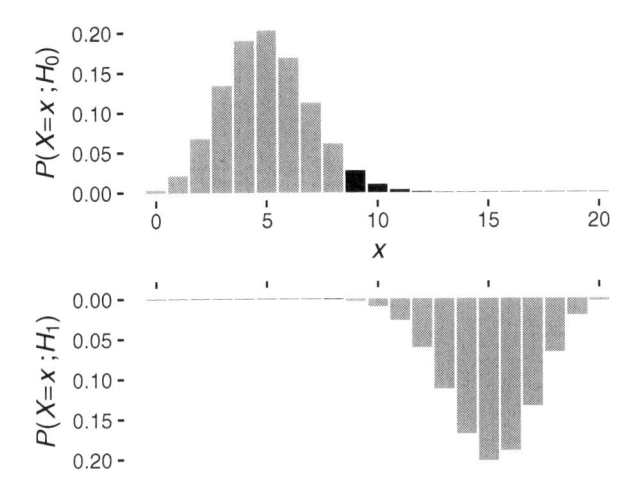

FIGURE 3.3 The likelihoods of the null H_0: $\theta = 0.25$ versus the alternative H_1: $\theta = 0.75$ when $n = 20$. The dark areas represent the type I and II error rates of the test, with the critical region set at more than 8 heads. That the two distributions do not overlap much means that the test has small error rates.

$9;H_0) \sim 0.04$, the significance level falls below 5% from $X \geq 9$, which suggests that we reject the null if more than 8 heads are observed. Calculating the probability of the type II error with this critical region, we obtain $\beta = P(X < 9;H_1) \sim 0.001$. Recalling that the same error rate of test B applied to 10 coin tosses is about 0.078, i.e., that test B overlooks a true alternative hypothesis 8 out of 100 times, we should regard this new error rate of one out of a thousand to be quite an improvement. From this we see that one can increase the confidence and power of a test by collecting more data. That is, a larger sample size makes us less prone to error when making a decision of rejection.

3.3 Philosophy of Classical Statistics

3.3.1 Testing as Inductive Behavior

Given the overview of the testing procedure just presented, let us now pause to think about how we should understand this procedure and its results from a philosophical viewpoint. Recall that the notion of the probability of a hypothesis does not make conceptual sense in frequentism. It thus follows that, in contrast to Bayesian inference, the job of a statistical test is not to decide which hypothesis is more probable. For instance, when a null hypothesis is rejected at a significance level of 5%, this does *not* mean that the probability of the hypothesis being true is less than 5%.[8] Such a statement is simply nonsense in the frequentist framework. Indeed, one may be disappointed to know that a statistical test by itself does not make any *direct* judgment about the truth or falsity of the hypothesis being tested (however, the qualification "direct" here will prove important later).[9] Then what does it do? The goal of testing theories is to provide systematic rules or algorithms for making judgments about probability models on the basis of data. In effect, a test is a function that maps data to a dichotomous choice between rejection or retainment of a null hypothesis: it returns a rejection if the data fall in the critical region, and a retainment otherwise. Based on this result, we make an actual judgment as to whether we should really reject the null, which then leads to certain concrete behavior (like buying the coin in the previous example). If this is how a test is used in our decision procedure, it goes without saying that the algorithm we depend on should be as reliable as possible. "Reliable" here means that the given test has a small risk of making mistakes: it has a high probability of rejecting a hypothesis when it is false and retaining it when it is true. These accuracy rates are nothing but the confidence and power that we saw earlier, which can therefore be taken as measures of a test's reliability. Note that the "probability" here is probability in the frequentist sense: it represents the relative frequencies or proportions of correct/incorrect answers when the test is conducted under similar situations over and over.[10] Specifically, the confidence and power indicate, respectively, what percentage of the verdicts of rejection and retainment made by the test in its repeated application are actually correct.

A statistical test, therefore, is a sort of diagnostic kit that returns a certain decision given some data, and testing theory measures and examines its reliability (Sober 2008). Thus, probability values like the significance level and the power assessed by such a theory are only *properties of the test* regarded as a diagnostic kit, and *not properties of the hypotheses* to which the test is applied (as in "probability of a hypothesis"), or of individual judgments resulting from applications of the test (as in "probability of this judgment being correct"). This is completely in line with the remark made in Section 3.1, namely, that probabilities for frequentists are properties of "collectives" *qua* types, so that it is a categorical mistake to think about probabilities (i.e., relative frequencies) of individual phenomena *qua* tokens. Hence, although the proposition "the probability of getting heads by tossing this coin repeatedly is one-half" makes sense according to the frequentist interpretation, the proposition "the probability of getting heads in the next toss of this coin is one-half" doesn't. Likewise, while it makes sense to ask about the relative frequency of correct answers in a long-run application of a given test, one cannot meaningfully ask about the accuracy rate of an individual decision resulting from this or that particular application of a test. This is why we stated previously that "a statistical test by itself does not make any direct judgment about the truth or falsity of the hypothesis being tested."

For these reasons, one of the founders of the theory of statistical testing, Jerzy Neyman, claimed that statistics does *not* provide a method of inductive reasoning, contrary to the expectation of many. Inductive reasoning, in his view, aims to evaluate the truth or falsity of a hypothesis on the basis of data. However, as we have seen, statistical tests do not make such judgments. What they proffer is a policy to guide our decisions under uncertainty. Because such decisions are part of our behavior, testing theory must be understood as a theory not of inductive reasoning, but of *inductive behavior*, or so argued Neyman (1957).

In fact, Neyman was not the first to emphasize the relationship between induction and behavior—a similar idea was espoused by David Hume, whom we encountered in Chapter 1. Recall that Hume denied the logical validity of inductive reasoning. Hence, if reasoning is to be understood as a valid derivation of a conclusion from premises, there is no such thing as inductive reasoning. This, however, does not prevent us from making inferences about the future from the past and acting accordingly. In effect, this kind of behavior is a *habit of mind*, which is engraved in us by repeated experience to form a code of conduct (Hume 1748). This habit governs our behavior in such a way that a certain experience, say of observing dark clouds, prompts us to perform a certain action, say of bringing an umbrella, even if we cannot theoretically prove that dark clouds bring about a storm. Hume called these habits "the great guide of human life" and thought they are formed automatically from experiences. Hence, although they are useful and indispensable in our everyday life, the rules that govern our inductive inferences are, like other habits, not something we form at our own will. But what if there were more than one of these guides and we could compare their accuracy rates? Tests are nothing but such guides or "habits" that govern our decisions about hypotheses on the basis of certain data (Hacking 1980). Each guide/test has its own characteristics and decision policy: some are cautious and do not reject the null unless there is sufficient evidence, while others put greater emphasis on making new discoveries at the expense of running some risk of false positives. In the face of such a variety of guides, the role of testing theory is to evaluate their performance and reliability in terms of error rates, α and β. In this way it serves as a "guide of guides," as it were, helping us to pick up the policy that best guides our decisions/actions.

3.3.2 Classical Statistics as Externalist Epistemology

According to Neyman, classical statistics is not about inferences but rather about policies for certain actions. Is this a satisfactory characterization of a statistical theory? That would depend on what we expect of statistical methods. The first widespread acceptance of testing theory was in the American military industry during World War II (Shibamura 2004). To ensure the quality of military supplies, the US army requested its suppliers to conduct sampling inspections and ship only those lots that contained fewer defective samples than a certain threshold. The problem for the suppliers was how to set the threshold. Setting the threshold too high would result in overlooking defective lots, thus undermining the trust of their customer; while if the threshold is set too stringently, the company would end up wasting many good lots and inflicting unnecessary costs. The suppliers thus needed a testing method that would minimize these risks in the long run, and classical testing theory provided an ideal solution for this purpose.

Be that as it may, statistical tests today are widely applied to scientific investigations, whose contexts and aims differ considerably from those involved in the quality control of mass production. Suppose, for instance, that medical research has reported that a null hypothesis that a certain new drug has no effect was rejected at a significance level of 1%. What do we expect from this result? Certainly not the "quality assurance" of this research group, that they will produce effective drugs in the long run. Rather, we are interested in whether the particular drug under question is effective or not. But according to Neyman's theory of inductive behavior, statistical tests do not tell us anything about the truth or falsity of a particular hypothesis. Does that mean that most applications of testing theory to scientific investigations rest on a gross misunderstanding of methodology and are therefore invalid? Again, at issue here is the concept of justification. By using a statistical test, scientists aim to justify a rejection or acceptance of a particular scientific hypothesis. But justification in what sense? More specifically, if testing theory is primary concerned with the long-term reliability of testing methods, how and in what sense is it capable of justifying individual judgments resulting from its applications? In what follows, we will try to fill this gap, again taking philosophical epistemology as a guiding thread.

Reliabilism

In the previous chapter, we defined knowledge as a true justified belief and introduced internalist epistemology as a strategy for justifying beliefs. Internalists think that a belief of an epistemic agent is justified when it is derived via a valid inference from other justified beliefs possessed by the same agent. This, however, is not the only method of justification. There is an alternative, *externalist* conception of justification. In the history of philosophical epistemology, the motivation for this view stemmed from a famous counterexample raised against the internalist concept of justification, called the *Gettier problem* (Gettier 1963). In this short but vastly influential paper, philosopher Edmund Gettier pointed out cases of apparently justified true belief, which, however, are not intuitively considered as knowledge. As an example, consider the following (true) story, which is adapted not from Gettier's own example but from an earlier discussion in Russell (1948). My office has a nice view of the university's clock tower (so I don't have a clock in my office). One day I looked at the clock after finishing my morning paper work, and it was pointing exactly at noon, so I went to the university cafeteria for lunch as usual. It was indeed 12 o'clock then, and students were just coming out of their classrooms after their morning classes. Coming back from the cafeteria, however, I found the clock hands still standing straight up. In fact, the clock was under maintenance that day, during which its hands were fixed to the same position. Now, the question we want to consider is whether I knew that it was noon when I went out for lunch. My belief that it was noon was, as it happened, a true belief. The problem then boils down to whether it was justified or not—but from the internalist perspective, my belief does seem to be justified by my visual perception of the clock tower. For one thing, it is my routine practice to check the time with the clock tower, and there is no reason to think that my perception on that particular day differed significantly from that on other days when the clock is operating normally. Hence, from the internalist perspective, the belief I formed on that day—that it is lunchtime—is a justified true belief, and thus qualifies as knowledge. I presume, however, that many of us would resist saying that I *knew* that it was lunchtime in this example. We would instead consider my having a true belief as a mere coincidence rather than knowledge. If so, the internalist definition of justification does not square well with our intuitive understanding of the concept.[11] Granted, a clash with common sense need not falsify a theory. Recall, however, that one of the motivations for justifying beliefs was to prevent "lucky guesses." Since the case in question in the preceding episode is precisely such a lucky guess, the Gettier problem at least points to the possibility that the internalist conception fails to fulfill this important function of justification.

If we deny that my belief in the clock tower story is one that is properly justified, perhaps the cause can be attributed to the way the belief was obtained. I gained the belief that it is noon by looking at a nonworking clock, rather than a fully functioning one. But an unmoving clock would hardly justify a judgment about time, for the simple reason that it is not a reliable source of information about time. This line of reasoning suggests the *reliabilist* conception of justification (Goldman 1975). According to reliabilism, whether a belief is justified or not is determined by the nature of the process that generated the belief. If the belief-generating process is reliable, in the sense that it produces more truths than falsities, then beliefs generated by such a process are justified. To understand this idea, suppose one of your friends believes that drinking three or more cups of coffee per day reduces the risk of stroke. Being suspicious, you ask her the source of that information. If she answers your question with a large-scale meta-analysis published in a prestigious academic journal, then you might think that her belief is justified, at least much more strongly than if her source were a shady internet article. If so, this must be because you think that the academic journal is reliable and reports true information more often than a random website. Conversely, an unmoving clock gives us the right time only for very limited moments in a day and hardly counts as a reliable source of information about time. For this reason, the belief about lunchtime I formed on that day from such a process was, even though actually true, not justified.

Compared with the internalism we considered in the previous chapter, the reliabilist concept of justification gives a markedly different answer to the philosophical question: what is justification? For internalists, justification is a matter of a relationship among beliefs possessed by the agent, or more specifically, whether a given belief is validly inferred from information accessible to the agent. Thus, all the factors that determine whether the belief is justified or not must reside within the agent (hence the name "internalism"). For reliabilists, on the other hand, the key factor in justification is the reliability of the belief-formation process, which is an objective fact not necessarily recognized by the epistemic agent. To see this last point, imagine that most papers published in the said high-profile academic journal were actually complete fakes with no reproducibility, fabricated solely by the hands of its chief editor. Then, any belief obtained from that journal, including your friend's belief about the beneficial effect of coffee, would not be justified according to the reliabilist criterion, even if no one except the guilty editor knew about the misconduct. That is, for reliabilists, whether a belief is justified or not is not determined by the subjective status of an epistemic agent, but depends in an important way on the objective situation holding outside the agent, such as the legitimacy of articles published in the journal, or the functionality of the clock. A view like this, which locates some of the justificatory factors outside of the epistemic agent, is called an *externalist epistemology*. Reliabilism is a kind of externalism, in the sense that it seeks the basis of justification in the objective reliability of the belief-generating process.

Nozick's Tracking Theory and Hypothesis Testing

Some questions still remain: what does it really mean to say that a certain epistemic process is reliable? And how do we assess this reliability? As a guide to answering these questions, let us refer to the view of another famous external epistemologist, Robert Nozick (Nozick 1981). Suppose an epistemic agent S believes that P. According to Nozick, an important condition that this belief must satisfy for it to count as knowledge is that the agent *tracks* the truth; that is, it must be the case that she comes to believe that P if P is indeed the case, and also that she would not believe that P otherwise. This tracking condition is expressed by the following pair of subjunctive conditionals:

(N1) If P were not true, S would not believe that P.
(N2) If P were true, S would believe that P.

Assuming that either P or not P is the case in reality, one of these conditionals describes a non-actual situation and thus is *counterfactual*. For this reason, Nozick's approach is sometimes called the *counterfactual theory of knowledge*, and hereafter we will call (N1) and (N2) *counterfactual conditionals*. If these two conditions hold, we can be assured that the agent S is forming her belief not haphazardly, but well in accordance with the actual state of affairs. Note that these tracking conditions can solve Gettier's problematic case shown earlier. In my clock tower example, I would still have believed that it was noon even if it actually weren't, because I was determining time based on a nonworking clock. That is, it was *not* the case that if it were not noon I would not have believed that it was noon. Nozick's proposal can thus rule out my belief on that day from counting as knowledge, since it fails to satisfy condition (N1).

Although the tracking theory was originally developed apart from the context of reliabilism, it is possible to combine them and read Nozick's two counterfactual statements as conditions for a reliable epistemic process. For it should be quite natural to demand that a given epistemic process, for it to be reliable at all, must track the truth in Nozick's sense, so that it judges that *P* just in the case that *P*, and not otherwise. Indeed, this seems to be exactly how we assess the reliability of epistemic processes, like our sensuous perceptions. Suppose, for example, that I am looking up at the blue sky now. I judge that the sky is blue on the basis of my epistemic process of vision. But if the sky were covered by clouds, or darkened by a solar eclipse, I would not have made that judgment. In this sense, my vision is a reliable epistemic process. On the other hand, suppose that I hear a note of C and judge that it is C. But since, unfortunately, I don't have a good ear, it is quite likely that I would have judged likewise even if the note I heard had been D or G. That is, my ear is not a reliable epistemic process (so don't try to take me out to karaoke).

We can summarize this discussion as follows:

> *Reliabilist-Tracking Justification*: Beliefs produced by a reliable epistemic process are justified. Reliable epistemic processes are those processes that effectively track the truth, by satisfying the conditions (N1) and (N2).

And then what? Well, the reason we have been discussing the externalist concept of justification so far is nothing but this: it bears out our intuition and expectation that statistical tests can be used to justify individual scientific hypotheses. To see this, first note that statistical tests that make a judgment about a hypothesis on the basis of data are a kind of belief-forming process. If, for instance, a particular test rejects the null hypothesis that a new drug is ineffective, we form the belief that this drug is indeed effective, and the opposite belief if the test fails to reject the null. Testing theory measures the reliability of this process in terms of the confidence coefficient and power of the test. Recall that the confidence coefficient of a test is the probability that it does not reject the null H_0 when it is true, or equivalently, the probability that it can detect the falsity of the alternative hypothesis H_1. Hence, if a test T has a high confidence coefficient, the following should hold:

(1) If H_1 were not true, T would not have accepted H_1.

Next, the power of a test is the probability of correctly rejecting a false null H_0, or equivalently, the probability of accepting the true alternative H_1. Hence, for a test T with a high power, we have:

(2) If H_1 were true, T would have accepted H_1.

In a nutshell, the confidence coefficient and power can be thought as indices that measure the extent to which a given statistical test satisfies Nozick's two counterfactual conditions of a reliable belief-forming process.[12] The higher these indices are, the more reliably the test "tracks" the state of affairs we are interested in, say, the efficacy of a newly developed drug. Thanks to Neyman and Pearson's testing theory, whether a given statistical test counts as a reliable belief-forming process in this sense can be evaluated in precise, probabilistic terms. It is in this reliabilist sense that a good statistical test can epistemically justify our beliefs and judgments about individual, specific scientific hypotheses, even though the goodness of a test is accessed in terms of long-term, frequentist criteria.

The Counterfactual Nature of Statistical Tests

We noted earlier the *counterfactual* nature of Nozick's two conditions. What does this mean? As the name suggests, counterfactual conditions ask us to imagine a possible circumstance that is not actually realized and to consider what would have happened if such a circumstance had obtained. The natural question, then, is how to determine the truth value of such sentences that purport to describe a counterfactual world. That is, we need to know the truth condition of counterfactual statements. Under which circumstances do Nozick's two conditions (N1) and (N2) become true or false? How do we know whether these circumstances hold or not?

The standard formulation of the truth conditions of counterfactual statements resorts to the concept of possible worlds. *Possible world semantics* asks us to imagine many possible worlds beyond the actual one in which we live. Let's take as an example the counterfactual statement "If I were a bird I would fly to you" and see how its truth condition is treated by possible world semantics. First, we need to imagine a set of possible worlds; in some of them I may be a bird, in another I may be a crocodile, in yet another I may not even exist; in some of them the earth is just like ours, in another it has four satellites, and in yet another it has almost no air like the moon, and so forth. Among these countless possibilities, focus on the set of worlds in which I am a bird but which in other respects are quite similar to the actual world. If in all such worlds I (as a bird) fly to you, then the counterfactual is true; otherwise—that is, if there is even one possible world in which I (as a bird) do not fly to you—the counterfactual is deemed false. The reason we restrict our attention to those worlds that are similar to the actual world (except that I am a bird) is because worlds that are too different are of little use in evaluating the counterfactual condition. For example, the earth in some possible worlds may be airless; then I would not be able to fly even if I were a bird, but such *force majeure* should not count as a breach of my pledge. On the other hand, if all the conditions are satisfied and yet I do not fly to you for some other reason, say, because I'm busy picking foods or flirting with other birds, then I would need to plead guilty to the charge of disloyalty. Hence, in *all* neighboring worlds, it must be the case that I fly to you. The truth value of Nozick's conditions (N1) and (N2) are to be assessed similarly. Taking the clock tower example again, we imagine possible worlds that are similar to the actual one except that P is not true, that is, worlds where I look up at the clock at some time other than noon, and then check whether I still believe that it is noon in such worlds. Since, by assumption, the clock is not functioning in those neighboring worlds, I would still believe that it is noon. We thus judge that the aforementioned counterfactual condition is not satisfied in this case.

Justification of statistical tests has similar, counterfactual characteristics (Mayo 2018). To begin with, a "hypothesis" in the frequentist framework is nothing but a statement about a possible world. In logic, a possible world is identified by a combination of propositions, namely, the set of propositions that are true in that world. In statistics, on the other hand, a possible state of the world is represented by a distributional family and its parameters. Specifying the distribution by a statistical hypothesis, therefore, amounts to identifying a possible world, with different specifications corresponding to different worlds. In one of these worlds, a phony shopkeeper is trying to rip you off with his bogus coin, while in another an honest dealer is offering a genuine error coin at a bargain price. The purpose of a test is to determine which among these possibilities is the actual world in which we live. Now, suppose you collect the necessary data and conduct a statistical test, which successfully rejects the null hypothesis. This result suggests that H_1 is true of the actual world we live in. But in order to justify this judgment, the counterfactual condition (N1) must hold, i.e., it must be the case that were H_1 not true, the test would not have rejected H_0. Under possible world semantics, this would be satisfied only if the test retains the null hypothesis H_0 in all the possible worlds similar to the actual one except for the fact that H_1 is false (hence H_0 is true) there. This, however, is too stringent a requirement for a stochastic hypothesis, for we may obtain unfavorable data merely by chance. Frequentists thus consider the question: in how many worlds, among those possible worlds in which the test is conducted, will the null hypothesis H_0 be rejected? This proportion is what we call the *p*-value. When this probability is so low that the erroneous rejection of H_0 is sufficiently rare, if not zero, among those possible worlds, the counterfactual condition is satisfied and the rejection of the hypothesis is justified. The case of non-rejection is treated similarly. That is, in such a case we imagine conducting the test in the possible worlds where H_1 is true and consider the probability of failing to reject the null hypothesis. This gives the power of the test, or the extent to which the test satisfies condition (N2), on the basis of which we can evaluate to what extent the test's conclusion is justified.

The salient feature of the counterfactual theory of justification is that according to the theory, whether a conclusion is justified depends crucially on neighboring possible worlds. In other words, whether a belief or judgment is justified or not cannot be determined solely by looking at the actual world. According to Nozick, whether an agent S's belief P is justified or not depends on counterfactual situations as to what S would have believed in the worlds where not-P. Likewise, whether a test T's rejection of the null hypothesis is justified or not depends on the counterfactual consideration of how frequently T would have rejected H_0 among all the non-H_1 possible worlds. Some may find it odd that the structure of possible worlds determines whether or not our belief or judgment in the actual world is justified, especially given that these different worlds are in principle utterly inaccessible and unknowable to us. There seems to be no way to actually check what I would do if I were to become a bird. Statistical tests make this apparent impossibility possible by introducing a theoretical structure into possible worlds. That is, it assumes that these possible worlds share the same statistical model (probabilistic kind) with the actual world, differing only in its parameters. This allows us to calculate the probabilities of hypothetical samples that should obtain in each of these worlds. We can then estimate the rate of rejection and retainment of a hypothesis among the possible worlds and determine whether the desired counterfactual conditions are satisfied or not. From this, one may say that frequentist statistics is a kind of statistics that probes the structure not of the actual world, but of possible worlds. Later, we will see that this counterfactual character takes on an essential significance in understanding the controversy between Bayesianism and frequentism.

3.3.3 Epistemic Problems of Frequentism

The Truth-Conduciveness of Statistical Tests

Recall that at the end of the previous chapter, having characterized Bayesianism as an internalist epistemology, we asked whether its concept of justification really possesses the property of truth-conduciveness that we expect of it, and if so, under what conditions. This problem stems from the coherentist nature of the Bayesian/internalist concept of justification: that is, if Bayesian justification is all about the logical coherence among beliefs possessed by an epistemic agent, it would need an additional account as to why these subjective beliefs could also be true, in the sense that they accord with objective facts. Does the same issue arise for frequentism *qua* externalist epistemology? That is, can we reasonably expect that a belief about a hypothesis justified by a reliable statistical test is actually true? From the reliabilist definition of justification we saw earlier, the answer seems to be yes. For reliabilists, to say that a belief is justified means nothing but that it was generated by a process that regularly produces beliefs in such a way that they accord with external facts. Hence the reliabilist concept of justification must be truth-conducive, trivially by definition.

Does this mean, then, that we can uncritically trust the results of a test with a high confidence coefficient and power? Indeed, it is common practice in scientific applications of statistical tests to take a low p-value as a straightforward statistical proof of the alternative hypothesis. Such an uncritical reliance on statistical significance in evaluating scientific hypotheses, however, has increasingly been viewed with suspicion, especially after the recent statement on the misuses of the p-value issued by the American Statistical Association (Wasserstein and Lazar 2016). The statement warns that the p-value, along with the underlying statistical theory and methods, is often misused or misinterpreted in the scientific community (the so-called *p-value problem*), leading to poor and unwarranted decision-making. Publications of research based on flawed statistical inferences often fail to replicate in subsequent experiments, giving rise to what is known as the *reproducibility crisis*. The suspicion has extended to the methodology itself, to the point that some scientific journals go so far as to ban the use of the p-value altogether.

Viewed through a philosophical lens, this recent revolt against testing theory has a striking similarity with a standard criticism made against externalist epistemology by the rival internalists. As mentioned previously, externalists do not require that an agent knows that the epistemic process she uses in forming her belief is reliable; it suffices that the process is reliable as a matter of fact, regardless of whether this fact is recognized by the agent. Internalists complain that this is too irresponsible and leads to several unintuitive consequences (Bonjour 1980). In particular, they claim that the externalist criterion ends up granting knowledge status to beliefs formed by an overtly unscientific, spooky method such as clairvoyance, as long as the method is reliable for some unknown reason, unintelligible even to the belief-forming agent. This problem indicates that an externalist knower, lacking a proper understanding of the process she uses, cannot rule out the possibility that the well-functioning of a supposedly reliable process is a sheer matter of luck. But if so, it fails to satisfy the very motivation of justifying beliefs, namely, to distinguish knowledge from mere "lucky guesses" (Chapter 2).

The question at stake in these two criticisms, one in statistics and the other in philosophy, is: to what extent, if at all, should one be responsible for the epistemic process one uses to acquire knowledge? In the eyes of internalists, externalists appear to abandon this responsibility altogether, and along with it the very essence of justification. Likewise, the ASA statement puts the blame on the thoughtless application of statistical tests unaccompanied by a proper understanding of the method or an inspection of their reliability, and claims that such irresponsible inferences fall far short of scientific justification. Classical statistics indeed has well-established testing protocols, but if one takes the reliability of tests as a given and pays little attention to why they can be trusted in the first place, don't they become something like mystical oracles?

Facing this challenge, one can think of two responses. One is to push externalist epistemology to its extreme and abdicate the responsibility of having a firm grip on the epistemological process one uses. This is roughly the strategy taken by some pragmatist-minded epistemologists such as Stich (1990), and in the next chapter, we will find its statistical counterpart in the recent development of deep learning, where increasingly greater emphasis is put on improving the performance of algorithms than on providing a theoretical warrant of their reliability.

The other, more moderate response takes the epistemic responsibility seriously and acknowledges the burden on the part of an epistemic agent to check the truth-conduciveness of the method in use. Such a stance would not take the reported outcome of a test for granted, but would require a second-order justification for the reliability of the decision process. From an epistemological perspective, this is tantamount to "internalizing" some external justificatory factors, thereby mitigating the externalist nature of classical statistics (Staley and Cobb 2011; Mayo 2018).

One concrete strategy along the second line is to verify or justify the assumptions underlying the testing setup. Statistical tests draw their appeal from the reliability of the epistemic process, which is supposedly measured by the p-value or power. The validity of these indices, however, depends on various implicit as well as explicit assumptions, most notably those concerning the likelihood function used to derive these indices. For instance, the construction and interpretation of the tests in our coin example rest entirely upon the distributional assumption, which remains untested over the course of the experiments. Fed with data, a statistical test issues a rejection or retainment of the null hypothesis, but it does not say anything about the validity of the test's assumptions, such as that 10 coin tosses can be represented by a binomial distribution, or that the probability of heads is either 0.25 or 0.75, and not any other value. Granted, coin tossing is an easy toy example with little room for doubt. But in other general cases, the question as to which probabilistic kind best represents the phenomenon of interest, or whether the target phenomenon can be represented by any definite probabilistic kind at all (that is, whether it really forms a well-behaved "natural kind"), poses a real empirical question whose answer cannot be taken for granted. Identifying an appropriate probabilistic kind and ensuring that the test setup actually conforms to that specification constitute an essential part of testing practice (at least parametric ones), upon which the validity of reliability measures like confidence coefficient and power crucially hinges. For this reason, a low p-value cannot be taken as a foolproof justification of a test's verdicts. We also need to ascertain whether the test itself is a properly functioning cognitive process, and satisfies the counterfactual tracking conditions we saw earlier. Proving this requires not just simply reporting a couple of indices produced by statistical software, but a critical scrutiny of the various assumptions that underlie these figures.

The preceding discussion concerned the condition under which a statistical test functions as a reliable belief-formation process. The reproducibility problem has an additional aspect, namely whether such processes are properly employed in scientific research. Any tool, if not used correctly, will not produce its intended effect. Likewise, improperly conducted statistical tests will not be able to justify hypotheses. Let us illustrate an inappropriate use of a statistical test with our familiar coin-tossing example. This time, imagine that the seller, instead of letting you throw the coin, only tells you that he tossed the coin ten times at his home and was able to reject the null hypothesis that this coin is not an error coin at the 5% significance level. Taking his words to be true, should you buy the coin? It all depends on how many times the seller conducted this experiment. It might indeed be true that he was able to reject the null hypothesis by tossing this coin. But it might also be the case that he conducted similar experiments with a lot of other coins he has at home. Then, even if these coins were all common ordinary coins, he could have rejected the null hypothesis with the probability of 5%—that is, 1 out of 20 coins—just by sheer chance. It may be the case that he simply picked out a coin that passed the test by plain luck, sweeping all the other unsuccessful coins under the carpet.

This kind of *multiple testing* can also be problematic in a scientific context. Imagine that a research laboratory tested the toxicity of 100 different chemical compounds and found that with 3 of them, labeled A, B, and C, the null hypothesis of being nontoxic was rejected at the significance level of 5%. But if the lab, based on this result, were to report the toxic risk of A, B, and C without mentioning the test results of the other 97 compounds, they would be committing the same deceit as the coin seller earlier. This deceit takes advantage of the stochastic nature of testing processes, namely, the fact that its reliability is guaranteed only probabilistically, so that repeated application of the test will produce a certain amount of false positives and negatives. "Hacking" a testing process in this way in to order to derive a desired conclusion is called *p-hacking*.

These considerations prompt us to rethink the externalist nature of frequentist epistemology. The externalist justification essentially relies on the processes or circumstances external to an epistemic agent. In my field of vision there now appears a blue sky through the window, which justifies my belief that it's a beautiful day today—or at least, so I think. The success or failure of this justification, however, depends on numerous external circumstances and conditions that have an influence on the functionality of my visual perception, such as that I don't have cyanopsia, or that there is no blue film on the window. These conditions are external to my visual process, and as such, they cannot be verified solely on the basis of the image it produces. Only when these external conditions are independently checked by other means can the justification by the cognitive process be guaranteed to be reliable, i.e., truth-conducive. Likewise, the externalist nature of frequentist justification implies that its truth-conduciveness is inherently dependent on the conditions external to the testing process. Testing theory formulates such external conditions through its assumptions on experimental design, the uniformity of nature (IID), probabilistic kinds (statistical models and likelihoods), and so on. However, whether or not these assumptions actually hold is ultimately "external" to statistical theory and eludes complete confirmation. At least, it is not something that can be read off from the results of the testing process. If that is the case, then judging the success or failure of a hypothesis solely on the basis of the test result or p-value is fundamentally misguided in light of the frequentist concept of justification. What really matters in the justification of a hypothesis in testing theory is the reliability of the process that produces it, not the particular result. The success or failure of the conditions that support this reliability is usually hidden from us, and is not something that can be confirmed by simple indicators. But in order for the externalist justification to be truth-conducive, we must take responsibility for these hidden external conditions and pay constant attention to their validity.

The Likelihood Principle

Attending to the epistemological nature of frequentist statistics also helps us understand the debate over the likelihood principle, a substantial theoretical criticism often leveled against frequentism by other statistical schools, such as Bayesianism. Briefly put, the *likelihood principle* is the claim that all information that matters to the inference of hypotheses and parameters is contained in the likelihood function for the observed data (Berger and Wolpert 1988). In other words, the principle states that the inferential implications that the data at hand have for a hypothesis depend only on the probability of obtaining that data under that hypothesis, and not on any other information. Hence, if two hypotheses have the same likelihood with respect to the observed data (or, more precisely, if they are proportional to each other by a constant factor), then the data should not give a preference between the two hypotheses. Conversely, if a hypothesis has the same likelihood under two different datasets, then these datasets should provide the same amount of evidential support for the hypothesis.

The standard Bayesian inference satisfies this principle, because by Bayes' theorem, the posterior probability $P(H|E)$ depends on the data only through the likelihood $P(E|H)$, provided that the same prior is used. For this reason, Bayesians adopt the likelihood principle as a fundamental principle that must be satisfied in any statistical inference.

On the other hand, the likelihood does not have the final say in frequentist inference. This is evident from the experiment discussed in Section 3.2.4, where a coin was tossed 20 times to determine if the null hypothesis $H_0: \theta = 0.25$

could be rejected. Recall that the critical region in this experiment is $X \geq 9$ if the significance level is to be set to 5% or lower. Hence, if the coin comes up heads 9 out of 20 times, we reject the null hypothesis H_0 and take it as evidence for the alternative hypothesis H_1: $\theta = 0.75$. But getting nine heads means that we had more tails; and this result by itself is more likely to be observed under the null hypothesis that the coin is biased toward tails than under the alternative hypothesis that it is biased toward heads, namely $P(X = 9;H_0) > P(X = 9;H_1)$.[13] Hence, if we focus only on the likelihood, the observed data seem to lend more support for the retainment of the null hypothesis than for its rejection. In fact, under the Bayesian framework, the posterior probabilities become $P(H_0 | X = 9) > P(H_1 | X = 9)$, provided that both hypotheses start with the same prior, so that the null hypothesis is more strongly confirmed. Statistical tests, in contrast, may reject the null hypothesis even if it has a higher likelihood than the alternative hypothesis, and this means that frequentist inference is influenced by information other than the likelihood.

In effect, frequentism does not satisfy the likelihood principle, and it has often been criticized by Bayesians and others in this regard. One of the "paradoxes" of frequentism that arises from its noncompliance to the likelihood principle is the *stopping rule problem* (Lindley and Phillips 1976; Howson and Urbach 2006; Sober 2008). This is a problem in which exactly the same data lead to conflicting conclusions in different experimental designs: rejection of the null hypothesis in one case and retainment in the other. We assumed previously that H_0: $\theta = 0.25$ and examined whether this could be rejected in a 20-coin-tossing experiment. Here, to fit the argument in Howson and Urbach (2006), let us take as the null hypothesis H_0: $\theta = 0.5$ and consider an experiment to see if the coin is unbiased. We take as the alternative hypothesis H_1: $\theta = 0.25$; that is, we assume that the coin is either fair or skewed to tails. In this case, it would be reasonable to reject the null hypothesis if the coin lands on heads significantly few times in 20 tosses. The plot on the left side of Figure 3.4 shows the probabilities of observing each number of heads when we toss a fair coin 20 times. Suppose we now perform this experiment and get 6 heads. Given that the probability of getting 6 or fewer heads under the null hypothesis is $P(X \leq 6;H_0) = 0.0577$, this result cannot reject the null (fair coin) hypothesis with the 5% significance level. However, this is not the only way to test the coin's bias. We can consider an alternative experimental design, where we keep tossing the coin until we observe 6 heads, at which point the experiment ends. This experiment is expected to drag on if the alternative hypothesis H_1: $\theta = 0.25$ is correct, but it will terminate relatively early if the null hypothesis H_0 is correct. We can thus denote by Y the total number of tosses it takes to get 6 heads (so naturally, $Y \geq 6$), and agree to reject the null if this value is greater than a certain value. This experiment looks a little complicated, but there is no logical flaw in it. That this experiment ends at the yth toss means that we get 5 heads (and $y - 1 - 5$ tails) by the $y - 1$-th toss and then another

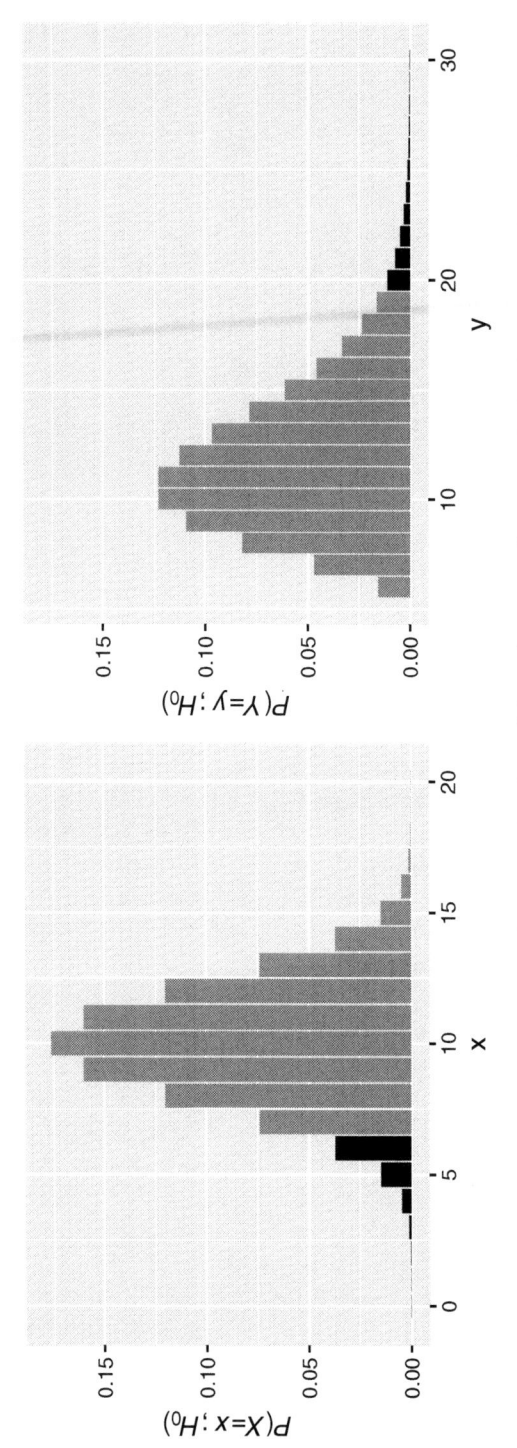

FIGURE 3.4 The probabilities of the experimental designs under the null hypothesis H_0: $\theta = 0.5$. Left: the *fixed design* makes judgments based on the number of heads one gets in 20 tosses. The probabilities of the outcomes fewer than 6 heads (shaded area) sum up to 0.0577. Right: the *flexible design* looks at the total number of tosses carried out until one gets six heads. The probabilities of carrying out 20 tosses or more sum up to 0.0318.

head again in the yth toss. Assuming the null hypothesis $H_0: \theta = 0.5$, the probability that this happens is

$$P(Y = y; H_0) = {}_{y-1}C_5 (0.5)^5 (0.5)^{y-6} (0.5)$$

where $y \geq 6$. The right figure in Figure 3.4 shows this probability, with the horizontal axis representing the number of times y that the coin was tossed until the end of the experiment. From this figure, we can calculate the type I error. Since we have decided to reject the null hypothesis when the number of tosses is above a certain threshold y', the probability of making a wrong decision is the sum of all the probability values above y' in this figure. Now we are assuming that we get 6 heads out of 20 tosses: hence, adding the probability values for 20 and more tosses, we obtain 0.0318, which is less than 0.05. Therefore, under this experimental design we *can* reject the null hypothesis at the 5% significance level.

Throughout this discussion, we have just one dataset: 20 tosses of a coin, out of which 6 are heads. This same data is seen as a basis for retaining the null hypothesis in one design, and rejecting it in the other. This, in the eyes of Bayesians, appears to reveal the arbitrariness and inconsistency of testing theory, as it shows that the result of the statistical analysis hinges on the scientist's choice of experimental design. Suppose, for example, that a researcher performs the same experiment with the intention of tossing a coin 20 times, and gets 6 heads as a result. Seeing that this result cannot reject the null hypothesis, she might conceal her original intention and pretend that she was actually conducting the experiment under the flexible design. She is then able to reject the null and justify the alternative hypothesis that the coin is not fair. Tying the fate of a hypothesis to the experimenter's whim like this should strike one as utterly absurd and inappropriate. In contrast, such a problem does not arise in Bayesian inference, whose posterior distribution is unaffected by the choice of stopping rules and depends only on the observed result; for this reason, Bayesians have argued that their method is less subjective in this regard (Howson and Urbach 2006, pp. 160, 248–250).

At first glance, this stopping rule problem may appear to be an artificial trick that exploits a "bug" in testing theory. But a closer look reveals that it in effect points to a philosophical difficulty that haunts any reliabilist epistemology, called the *generality problem*. Reliabilists seek the source of justification in the process that produces beliefs. A given process, however, can be described at various levels of granularity and detail. For example, imagine that someone takes a pill and tells you that the pill is tasteless. You may take her judgment to be generated by "the gustatory process of a human with no taste disorder," which should be reliable to some extent. If, however, she happens to have just eaten a very hot curry, this process is also "the gustatory process of a human with no taste disorder after having eaten a stimulant," which does not sound very reliable. Alternatively, you could take a very broad view and think of the process as "the process by which a multicellular organism recognizes an external substance through its stimulus receptor cells," in which case you would certainly not regard it as a reliable indicator of taste. In this way, the same event can be described as the result of many (not numerically, but descriptively) different "processes." The reliability of which among these processes, then, should be taken as the standard in justifying a given belief? This is the generality problem (Conee and Feldman 1998). Simply increasing the granularity of the process is no solution, for that will eventually leave us with the most specific description applicable only to the single instance at hand, thereby making it impossible to estimate its reliability or to express it in terms of frequentist probability (see Section 3.1). The reader might note that this problem is similar to the reference class problem we encountered in Section 2.3.3. The problem is a genuine one with no universally accepted solution, and thus it has been considered a major challenge to reliabilism.

Regarding a statistical test as a cognitive process, we see that the stopping rule problem is none other than a version of this general epistemological problem. An important moral of the generality problem is that one and the same experiment may be described as different statistical tests that give rise to conflicting conclusions, and one cannot determine which among them should be used just by looking from the outside. The previous example can be described either as a process of tossing a coin 20 times, or that of tossing it until one gets 6 heads. Depending on the description we take, the belief that the coin is biased may or may not be justified, and there is no principled way to decide which process is the one actually being used. But how, then, can testing theory serve to justify any hypothesis?

One frequentist response to this challenge is to bite the bullet and fully acknowledge the indeterminacy of processes, while stressing that it is actually a virtue of their theory that it recognizes this indeterminacy. Mayo (1996), for instance, emphasizes that data should always be interpreted in accordance with the experimental design; hence, on her view, the blame is on the side of the Bayesians who ignore this dependency and pretend as if a hypothesis can be accepted or rejected based just on the data. This response makes sense in light of the reliabilist nature of frequentism. If a statistical test is a sort of cognitive process for drawing conclusions from data, there should be no contradiction in there being multiple such processes which come to conflicting conclusions in the face of the same data. What is important is that the experimenter, recognizing this fact, makes it explicit which process he or she uses in the experiment and strictly follows that decision. The aforementioned case of switching from the originally planned design to the alternative upon looking at the data violates this rule and is hence prohibited. At fault here is not so much the testing methodology itself as the deviation from its protocol, which requires the experimenter to prespecify the cognitive process or method she will use during the experiment. The prior specification is necessary because the frequentist justification of hypotheses inherently relies on the reliability of the process, so that its decision of rejection or retainment is justified only insofar as it comes from the correct application of a reliable cognitive process. From this perspective, using a testing method that allows one to draw a desired conclusion come what may hardly counts as reliable, for it would not track the truth in Nozick's sense. At any rate, the reliability of such an opportunistic "process," if even it can be estimated at all, falls far short of that indicated by the confidence coefficient and power of the individual testing processes. Such a practice is thus a "hacking" of the testing process, just like the multiple testing discussed earlier, for it amounts to presenting the reliability measure of a testing process which in fact was not used in the actual experiment.

It is not just the data, but rather the process used to derive the conclusion that matters in the frequentist justification. This implies a sort of relativism, to the effect that any conclusion of rejection or retention of the null hypothesis is justified only in light of a particular experimental design and other testing conditions, and not unconditionally. A conclusion justified by one process may not be so by another—to embrace the generality problem is to admit that there is no unique, ultimate justification that would clear up all such relativism. It follows that scientific conclusions reached through statistical tests can be justified only relative to the method used. This does not necessarily mean "anything goes," since we can explicitly assess the reliability of each testing process with the aid of testing theory. Doing so, however, requires more than a superficial look at the result or p-value of a statistical test: it demands a careful consideration of the overall experimental design and testing method, including the experimenter's intentions.

What lies underneath the difference between frequentist and Bayesian attitudes toward the likelihood principle is a difference in their epistemology. For Bayesians and internal epistemologists, data are the sole foundation upon which our empirical reasoning stands, the epistemological "given," and must contain everything we will ever be able to know about the world. If so, everything that can be inferred about a statistical model must be summarized in the likelihood, i.e., the probability of obtaining the data under that model. For externalism, on the other hand, data do not mean everything. What matters is rather the external circumstances in and with which the data were obtained, as well as the reliability of this acquisition process. Drawing on Nozick's argument, we characterized this reliability in terms of two counterfactual conditions and mapped each to the confidence coefficient and power of a statistical test. This means that the frequentist justification process attaches weight not only to the actual data, but also to the "counterfactual data," so to speak; namely, the data that were not actually observed but could have been observed. This is because to consider what the test would conclude should the null hypothesis be true or false amounts to contemplating what kind of data would be obtained in a world that is (possibly) different from the actual one. Such counterfactual information is obviously not revealed by observation, and for this reason frequentists resist the likelihood principle, which demands that we ground all our inferences on the *actually* observed data.[14]

3.3.4 Summary: Beyond the Bayesian vs. Frequentist War

In this chapter we examined the frequentist methodology and its problems from an epistemological perspective, with a particular focus on the theory of statistical testing. Unlike Bayesianism/internalism, which locates the justification of probabilistic inferences in the logical consistency among beliefs, frequentism/externalism seeks the basis of justification in the reliability of the inferential process, of which statistical tests are a kind. The challenge for frequentist epistemology, then, becomes how to ensure the reliability of such external processes. As we have seen throughout this chapter, classical statistics helps us to assess the reliability of inferential processes on the basis of assumptions about probabilistic kinds and experimental design. These assumptions, however, stand "external" to the inferential process, and there is no principled way to evaluate their validity from the data in a systematic fashion.

Bayesians with an internalist slant will see a serious problem here. In their eyes, the assumption of a probabilistic kind is no more than a belief or "doxa" that an agent projects onto the phenomenon under study. If so, there should be a certain qualification on such beliefs, and this qualification should be reflected in the subsequent statistical analysis in the form of prior probability. Indeed, by admitting prior probabilities of hypotheses, Bayesians seem to take a more cautious and flexible attitude toward the assumption of probabilistic kinds. This is thanks to the fact that Bayesian statistics reduces all uncertainties to internal "degrees of beliefs" and treats them uniformly under the principle of Bayes' theorem.

Frequentists, on their side, retort that Bayesians' obsession with this kind of uniform treatment makes them overlook important factors in inference. They emphasize that inference should depend not only on the assumption of a probabilistic kind but also on a myriad of external factors, such as experimental design and stopping rules, which cannot be summarized by or reduced to just the likelihood and prior probability. It is thus the Bayesians who, clinging to the likelihood principle, turn their eyes away from essential factors in inductive inference, or so the frequentists argue.

The purpose of this book is neither to settle this dispute nor to add fuel to it. What, then, is the point of pondering about these "isms"? The main reason we have to pay attention to these philosophical foundations is that inductive reasoning, by its nature, involves muddy uncertainties not amenable to clearcut logical or mathematical treatment. The crux of inductive reasoning lies in inferring what we don't know from what we do know; but making such nondeductive inferences in a logically valid way is essentially impossible, as Hume had already perceived in the 18th century. The "Humean predicament is the human predicament," Quine once remarked, and this is still so even with today's mathematically sophisticated statistical theory. In order to fill the inevitable logical gap between the premises and conclusions of inductive reasoning, inferential statistics introduces certain ontological posits such as uniformity and probabilistic kinds, and explores their nature with the lead of the given evidence. The lead is only partial and indirect, for these entities transcend experience and elude direct access. This inevitably raises the epistemological question as to how and in what sense statistical inferences to inherently unobservable entities can be justified. The previous and current chapters tackled this question by focusing on two kinds of concepts of justification, internalist and externalist, and by connecting them to Bayesian and frequentist statistics, respectively. Our analysis, if successful, revealed that the two camps differ not just in their methodology but also at a deeper conceptual level, namely, regarding what it means for a hypothesis to be justified by statistical analysis at all. If so, we cannot argue which is better without looking at this fundamental difference. From this perspective, the oft-raised issues concerning the arbitrariness of prior distributions or violation of the likelihood principle are only surface symptoms of the deeper conceptual division, like two floating icebergs colliding at their tips. In order to fully appreciate the motivation of the mutual criticisms and move the discussion in a constructive direction, we need to direct our gaze to the submerged philosophical ideas.

Throughout this book we have characterized inferential statistics as a means for justifying scientific hypotheses, and emphasized that to fulfill this role it must be truth-conducive—i.e., there must be a reason to believe that its conclusions put us on the right track to uncovering the target phenomenon. This original motivation of statistical analysis, however, sometimes falls into the background in practice. This is particularly true when one feeds collected data into a packaged software to obtain a one-click statistical solution without questioning its theoretical underpinnings. Conclusions from such a "recipe-like statistics" (Mayo 2018) hardly provide *bona fide*, truth-conducive justifications. Calculating posterior distributions or obtaining low *p*-values, by themselves, do not tell us anything about the target phenomena. Bringing such results into the service of inference-making about the real world requires more than following a routine protocol and internal logic of a given statistical method; it further demands that we check whether the justificatory procedure in use is actually truth-conducive. What, then, makes statistical methods truth-conducive? In the current and previous chapters, we attempted to draw an answer to these questions from their respective epistemological natures and point to conditions the methods must satisfy in order to fulfill their original motivation.

Of course, philosophical reflection is not a silver bullet. It does not fill the logical lacuna in inductive reasoning, nor can one epistemological standpoint, either internalism or externalism, offer a complete picture of statistical justification. In the previous chapter we asked how Bayesian epistemology, which focuses primarily on the logical coherence of beliefs, can ensure correspondence with external facts; and we suggested that to overcome this difficulty, one must step out of the internalist framework. In this chapter, we took up the p-value problem and the reproducibility crisis as problems raised against frequentism and argued that these problems highlight the need to take responsibility of the truth-conduciveness of the statistical testing process. This requires frequentists to step back from pure externalism and to some extent "internalize" its justificatory machinery. The moral we should draw from these apparently unrelated problems is that any statistical method, either Bayesian or frequentist, cannot be content with staying within just one epistemological framework. Epistemology does not vindicate any particular statistical method or the concept of justification used therein. If anything, it is unlikely that epistemologists will ever agree on what concept of justification is the "correct" one, or even that the question has a definitive answer. Even so, this meta-statistical question is unavoidable, given that we have to rely on some justificatory concept or another if we are to justify a scientific hypothesis using a statistical method. The least we can do, then, is to be aware of the conceptual framework we adopt in our reasoning. Philosophical analysis sheds light on these ideological backgrounds and conundrums that are often forgotten in the practice of statistical methods, and provides a meta-perspective for comparing different methods and understanding the debates that take place between them.

Further Reading

For philosophical expositions of the frequentist interpretation of probability, see the aforementioned Gillies (2000), Childers (2013), and Rowbottom (2015). The basics of statistical testing are covered by almost any statistics textbook, including Vaughan (2013) and Wasserman (2004) mentioned in Chapter 1. Gillard (2020) also gives a succinct account. Deborah Mayo (1996, 2018) is a leading advocate of the frequentist philosophy. Externalist epistemology, along with Nozick's tracking theory, is briefly discussed in Nagel (2014).

Notes

1. Recall that we defined $H = 1$ as the inverse image of a random variable H, i.e., as the subset $\{\omega \in \Omega : H(\omega) = 1\}$ of Ω.
2. However, this argument does not straightforwardly carry over to complete or σ-additivity, i.e., cases in which there are infinitely many terms. In fact, relative frequency by itself is not countably additive. See van Fraassen (1977) and Gillies (2000) for details.

3. "The phrase 'probability of death', when it refers to a single person, has no meaning at all for us" (von Mises 1928). Dissatisfied by this conclusion, some, including most notably Karl Popper, proposed that a single-case probability can be defined as its *propensity* on the basis of a long-term frequency. For more details, see Gillies (2000) and Suárez (2020).
4. Granted, a probability function, as a mathematical measure, assigns a value to each individual element of the sample space. Such an assignment, however, is only a mathematical construct which by itself does not have any realistic/objective meaning in frequentism. A probability value has a counterpart in the actual world only in light of a certain probability distribution and collective. In this respect, one may say that in the frequentist interpretation, probability functions are ontologically prior but semantically posterior to probability distributions.
5. As we saw in the discussion of the base rate fallacy, even in the subjectivist framework one cannot conclude that the hypothesis is improbable.
6. In this book we deal only with the most elementary kind of test, namely, a test between two simple hypotheses H_0 and H_1, each denoting a specific distribution or parameter. A hypothesis is called composite when it ranges over multiple distributions or parameters (as when one tests whether the mean is zero or not, in which case the alternative H_1 consists of infinitely many parameter hypotheses $\{\mu | \mu \neq 0\}$).
7. In Bayesian statistics, the likelihood is the conditional probability $P(X = x | H_0)$ given the hypothesis. But, for the reasons discussed in the previous section, in the frequentist framework the hypotheses H_0, H_1 are neither random variables nor values thereof, which means that we cannot condition probabilities on them. For frequentists, hypotheses are expressed by their respective probability distributions, and thus, to be precise, the notation $P(X = x; H_0)$ stands for "the distribution of X resulting from the probability function P_{H_0} indexed by H_0."
8. Note, however, that even in Bayesian statistics, what we can calculate is *a degree of belief* in a hypothesis, and not *the probability of that hypothesis being true*, whatever that may be.
9. "As far as a particular hypothesis is concerned, no test based upon the theory of probability can by itself provide any valuable evidence of the truth or falsehood of that hypothesis" (Neyman and Pearson 1933, p. 291).
10. In effect, the result of a given test is a function of the data understood as random variables, and is thus itself a binary random variable which has its own probability distribution. Significance level (or confidence coefficient) and power are parameters of this distribution. If the distribution of the original data is to be interpreted in terms of frequentist probability, so must these parameters.
11. Another response would be to keep the internalist concept of justification and take the Gettier case as a counterexample to the traditional definition of knowledge as a true justified belief. Although it is popular to interpret the Gettier problem in this way, in this book we keep the traditional definition of knowledge as unquestioned and focus on the concept of justification.
12. This connection has been mentioned briefly by Nozick himself (Nozick 1981, p. 260). More recently, Roush (2005) has offered a Bayesian interpretation of the tracking theory. In my view, however, the counterfactual nature of the tracking theory is best captured by classical, rather than Bayesian, statistics.
13. As a matter of fact, we have $P(X = 9; H_0) \approx 0.027$ and $P(X = 9; H_1) \approx 0.003$, with the former being about one order of magnitude larger.

14. It is known that the likelihood principle is equivalent to a conjunction of two other principles: the principle of sufficiency, according to which statistical inferences are fully determined by sufficient statistics, which summarize the whole data; and the principle of conditionality, which states that inferences should be based only on experiments that were actually done, and not on possible experiments that could have been performed but were not in reality (Birnbaum 1962). The discussion here suggests that frequentists, by resorting to counterfactual information, do not accept the principle of conditionality.

4
MODEL SELECTION AND MACHINE LEARNING

In the previous two chapters, we considered how traditional statistics—both Bayesian and classical—has dealt with the problem of induction. In order to predict the future from the past, one has to assume what Hume called the "uniformity of nature" that remains invariant across time. In inferential statistics, this uniformity is represented by a probability model and is estimated from observed data. The estimated model then allows us to indirectly predict future samples, which we suppose are generated from the same model (see Figure 1.2). On this picture, one may naturally be led to expect that a better and more precise determination of the underlying uniformity or probability model should allow for better predictions and inferences. Indeed, this was one of our motivations for our exposition in the previous two chapters on Bayesian and classical statistics as empirical methods for identifying a probability model *qua* model of reality. This presumption, however, may be questioned. For the purpose of prediction, is it always a good idea to strive to delineate the data-generating process as accurately as possible? Isn't it possible that statistical models that are not so faithful to nature are able to make better predictions? Surprisingly, the answer is yes. It is not always the case that precise models make better predictions; sometimes models that "distort" reality a bit fare better. In this chapter, we will see this through two cases, model selection and deep learning.

4.1 The Maximum Likelihood Method and Model Fitting

Before diving in, let us warm up and familiarize ourselves with the notion of model fitting and its representative technique, *the method of maximum likelihood*, which will appear frequently in this chapter. The "models" we will discuss here

DOI: 10.4324/9781003319061-5

are nothing but statistical models, or what we dubbed "probabilistic kinds" in Chapter 1. They are, recall, hypotheses about distributions of given random variables, represented by a distributional family with a (set of) parameter(s). We have seen, for instance, that the number X of heads in 10 coin tosses can be modeled by the binomial distribution $P(X; \theta)$, with the parameter θ representing the probability of heads. As in the previous chapter, the semicolon (";") conveys the idea that the distribution of X is determined by the parameter θ. (See also note 7 of Chapter 3. In the Bayesian framework, $P(X; \theta)$ can be read as a simple conditional probability, $P(X|\theta)$).

If we were to follow the lead of the previous two chapters, we would then proceed to estimate these parameters, and for this purpose we would need to make some hypotheses about their values. Bayesians would calculate a posterior distribution from an assumed prior distribution of θ, while frequentists would make a particular hypothesis (i.e., null hypothesis) about the value of θ and test its validity. But here let us take a different approach and simply consider fitting the parameters to the data at hand, without going into the business of confirmation. That is, we set aside the question of what the real probability model looks like, and instead ask for a set of parameter values that would best "predict" the observed data.[1] In other words, our goal here is to find the parameter values that maximize the model's likelihood, i.e., the probability of obtaining the data under a given probabilistic kind.

So let's try it. Suppose that you flip a coin and get 6 heads out of 10 tosses ($X = 6$). Which among the possible parameter values $0 \leq \theta \leq 1$ makes this outcome most likely, or in other words, maximizes the likelihood $P(X = 6; \theta)$? From the binomial distribution, the likelihood $P(X = 6; \theta)$ is given by:

$$_{10}C_6 \theta^6 (1-\theta)^4. \tag{4.1}$$

This is a function of θ, and is illustrated by the plot in Figure 4.1. The graph appears to hit the highest likelihood around $\theta = 0.6$. This can be confirmed by

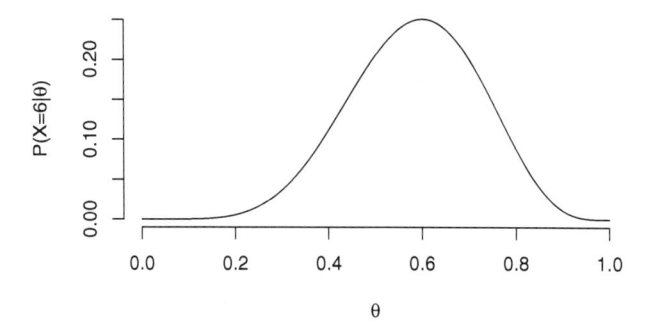

FIGURE 4.1 The plot of the likelihood of the binomial model with parameter $0 \leq \theta \leq 1$, when 6 heads are observed out of 10 tosses.

differentiating likelihood function (Equation 4.1). Recall that a derivative of a function gives the slope of its graph at a given point. Since the slope becomes zero at the vertex, to maximize likelihood function (Equation 4.1), one can just differentiate it with respect to θ and set the result to zero. The differentiation gives

$$\theta^5(1-\theta)^3(6-10\theta) \tag{4.2}$$

(we ignored $_{10}C_6$ since it is a constant that does not involve the parameter). This becomes zero if $\theta = 0$, 1, or 0.6. Seeing that the graph hits the bottom instead of the top at 0 and 1, we conclude that the likelihood is indeed maximized at $\theta = 0.6$; that is, the hypothesis that best accommodates the result of 6 heads out of 10 tosses is that the coin lands heads with probability 0.6.

In Figure 4.1, we took as our target statistical model the binomial distribution, which has only one parameter θ. In a similar way, when a model has multiple parameters, one may seek for the tuple of parameters that maximizes the model's likelihood. The parameters thus obtained are called *maximum likelihood estimators*, or MLE for short, and denoted by $\hat{\theta}$ with a hat. We denote the maximized likelihood of the model M by $\ell(M)$.[2] In this example, we obtained the maximum likelihood of the binomial model by maximizing its likelihood function (Equation 4.1). But in practice, one usually tries to maximize not the likelihood function itself but its logarithm, called the *log likelihood*. This is only for the sake of computational convenience (by taking logarithms, products of probabilities become sums, which are much easier to compute), and does not alter the result—that is, the parameters that maximize the likelihood also maximize the log likelihood, and vice versa. A model's maximum log likelihood will be denoted by $\log\ell(M)$.

As long as our target model is simple as in the previous example, its maximum (log) likelihood can be easily obtained by differentiating the likelihood function. But when the model is complex and has a number of parameters, its likelihood function also tends to get complex, making it difficult to find its maximum analytically. Even if the likelihood function is differentiable, it may be exceedingly hard to solve the relevant extremum problem. In that case one needs to approach the peak by climbing the likelihood function step by step. Let us illustrate this procedure with the aforementioned binomial case. (Since this particular problem can be solved analytically, as already shown, this procedure is unnecessary. We do this only for the purpose of illustration.) First, we pick a departure point at random. Let us start from, say, $\theta = 0.4$. Plugging this into the derivative (Equation 4.2) of the likelihood function gives us the slope at this point. The result is $(0.4)^5(0.6)^3(2) \sim 0.004$, which means that the slope is positive and rising toward the right. So we climb to the right a little bit, say to $\theta = 0.5$, and then calculate the slope at this new point. Repeating this procedure will bring us to the peak of the likelihood function, $\theta = 0.6$.

Such a straightforward hike to the peak, however, is guaranteed only for "Mt. Fuji-shaped" likelihood functions with a single peak; for more complicated functions with rugged landscapes, the step-by-step climbing will likely to bring you to a local optimum/peak, unless you are lucky enough to start from the foot of the global optimum. The climbing in our example was further simplified by the fact that it takes place in just one dimension (along the θ axis), because the model had just one parameter. But if it had $n \geq 2$ parameters, the mountain to be conquered is an n-dimensional hypersurface, and we need to check the slope along n directions at each step, making the numerical search for the maximum likelihood estimator much more cumbersome. We will return to this issue later when we discuss deep learning.

The method of maximum likelihood thus tries to find the parameter values of a model that best accommodate or "predict" observed data. Adjusting a model to a particular dataset is called *model fitting*, and an adjusted model is called a *fitted model*. The maximum likelihood method is a popular model-fitting technique, but not the only one. Another famous approach is the *least squares method*, which aims to find the parameters that minimize the discrepancy of a model's predictions with the actual values. Either way, the goal is simply to fit a model to a particular dataset, nothing more or less. In particular, the maximum likelihood and other model-fitting methods do not imply anything about the correctness of the model—i.e., there is no connotation that the fitted model likely captures or approximates reality. This is obvious from our example: even if the maximum likelihood estimator after 6 heads is $\hat{\theta} = 0.6$, it would be premature to conclude that the coin is biased toward heads.[3] The maximum likelihood method just picks the hypothesis that best fits this or that particular dataset, without paying attention to whether that hypothesis is true in a more general setup. In this respect, the maximum likelihood and other model-fitting methods have a different purpose and nature from those of the Bayesian estimation and hypothesis testing methods that we saw in the previous chapters.

4.2 Model Selection

4.2.1 Regression Models and the Motivation for Model Selection

With these remarks in place, let's move on to the main topic of this chapter, prediction. To make things concrete, we consider a simple regression model as a means for prediction. In general, *regression* is a method where one uses a set of variables X to predict or classify the value of another variable Y.[4] We encountered regression when we introduced Galton in Chapter 1, but within the confines of descriptive statistics, its use was limited to summarizing past data. Our goal here, in contrast, is to predict the unobserved on the basis of the observed, and that falls into the realm of inferential statistics. If we were to use Galton's data to infer the height of unobserved families who lived in London in that era, then our problem would be that of prediction, and we would use regression for that purpose. Such regression problems are ubiquitous: inferring weight from height, predicting college admission from SAT scores, and detecting an object, say a cat, in visual images are typical examples.

In regression models, the variables that serve as the basis for prediction are called *explanatory variables*, while those that are to be predicted are called *response variables*.[5] In the three examples just described, the explanatory variables are height, SAT scores, and images, while the response variables are weight, college admission, and "cathood," respectively. Of course, we could investigate more than one explanatory variable; for instance, we could take into consideration not just the SAT scores but also other factors, like essay scores, high school grades, and so forth in predicting college admission. In this case, the set of explanatory variables is expressed by a vector $\boldsymbol{X} = (X_1, X_2, \ldots, X_n)$, with values $\boldsymbol{x} = (x_1, x_2, \ldots, x_n)$. There should also be a host of other factors that are not explicitly registered as explanatory variables. These factors are lumped together as an error term, expressed by ϵ. An error term ϵ is a random variable that is assumed to follow a particular distribution. With this setup, a regression model describes the response variable as a function of the explanatory variables and error term:

$$y = f(\boldsymbol{x}, \epsilon).$$

The aim of a regression problem is to determine the functional form of f which would allow good predictions of y from input $\boldsymbol{x}$. This is done by first determining the general form of the function f and then working out the details by adjusting its parameters. That is, we again introduce a "probabilistic kind," as we did in Chapter 1. Indeed, what people call a (parametric) regression model is nothing but a probabilistic kind. Such probabilistic kinds/models come in various guises, but the simplest *linear regression model* has the following form:

$$y = f(\boldsymbol{x}, \epsilon; \boldsymbol{\theta})$$
$$= \beta_1 x_1 + \beta_2 x_2 + \cdots + \beta_n x_n + \epsilon$$

which expresses the response variable as a sum of the explanatory variables and error term. The parameters $\boldsymbol{\theta}$ of this model are the regression coefficients β_1, $\beta_2, \cdots$ which measure the relative importance of each explanatory factor, and the parameters that determine the distribution of the random error term ϵ. For instance, if the error term follows a normal distribution, the parameters will be the mean μ and variance σ^2. In this case, the mean μ determines the y-intercept of the regression line, while the variance σ^2 represents the dispersion of data around the line. The first line of the aforementioned formula, $f(\boldsymbol{x}, \epsilon; \boldsymbol{\theta})$, expresses the fact that the regression model f is fully specified by the parameters $\boldsymbol{\theta}$, and that through this function the distribution of the response variable is determined when a particular input x is given.

Since a regression model is nothing but a parametric statistical model, or a probabilistic kind in our parlance, we may resort to the traditional inferential statistical approach we saw in the previous chapters to estimate its parameters, and then use it for prediction. In the Bayesian approach, one can calculate the posterior distribution for each parameter from the data and then derive the posterior predictive distribution (Section 1.2.3). Frequentists, on the other hand, can test whether each regression coefficient significantly differs from zero, or calculate their confidence interval (not covered in this book). However, things get much simpler when one uses the maximum likelihood method. In this case, one seeks the $\hat{\theta} = (\hat{\beta}_1, \cdots, \hat{\mu}, \hat{\sigma}^2)$ that maximizes the probability of observed data. As we saw in Section 2.3.2, once the parameters of a probabilistic kind are fixed, so is the joint distribution of X and Y, and thus the probabilities of their values can be calculated; one can then use these probabilities to predict Y.[6]

All of these procedures assume *a particular probabilistic kind/regression model* as the data-generating process, and aim to correctly identify that probabilistic kind for the purpose of prediction. In the previous example, for instance, we adopted a linear model that takes X_1 to X_n as the explanatory variables to be used for prediction. But how should we select a particular model to begin with? In most cases there are several candidates. For predicting college admission, one may consider a model that uses only SAT scores, or those that also include essay scores, high school grades, and so on. Which out of these various candidates should be selected? The theory of *model selection* aims to answer this question. As we will see, it provides a criterion for choosing one among multiple probabilistic kinds or models on the basis of their predictive performance.

4.2.2 A Model's Likelihood and Overfitting

In this section we explain the idea of model selection using a simple example, with a particular focus on the theory of Akaike's information criterion (Akaike 1974). Consider two linear regression models

$$M_1 : y = \beta_1 x_1 + \epsilon, \qquad \epsilon \sim N\left(\mu_1, \sigma_1^2\right) \tag{4.3}$$

and

$$M_2 : y = \beta_1 x_1 + \beta_2 x_2 + \epsilon, \qquad \epsilon \sim N\left(\mu_2, \sigma_2^2\right) \tag{4.4}$$

where $\epsilon \sim N(\mu, \sigma^2)$ means that the error term ϵ follows the normal distribution with mean μ and variance σ^2. Let us denote the parameters of the first model by $\theta_1 = \left(\beta_1, \mu_1, \sigma_1^2\right)$ and those of the second model by $\theta_2 = \left(\beta_1, \beta_2, \mu_2, \sigma_2^2\right)$.

More specifically, let's interpret these regression models as models for predicting the GPA score Y of college students, where M_1 takes only the SAT score X_1 as the explanatory variable, while M_2 also takes high school grades X_2 into account.

How can we decide which of these to use for prediction? An initial idea may be to see how these models fit the data—that is, to compare their maximum likelihood $\ell(M_1)$ and $\ell(M_2)$. If we have data $d = (x_1, x_2, y)$, we can apply the maximum likelihood method to obtain MLE $\hat{\theta}_1, \hat{\theta}_2$, which respectively maximize the likelihood of the models M_1 and M_2. This allows us to compare their maximum likelihoods

$$\ell(M_1) = P(d; \hat{\theta}_1), \qquad \ell(M_2) = P(d; \hat{\theta}_2)$$

Since the likelihood measures how well a model accommodates data, it seems reasonable to choose the model with the higher likelihood as the better one.

This strategy, however, does not work. To begin with, no matter what the data is, M_1's likelihood never exceeds that of M_2, so it is always the case that $\ell(M_1) \leq \ell(M_2)$, making the attempted "comparison" meaningless. This is because M_1 is just a special case of M_2, without x_2. Hence, however well M_1 fits the given data, M_2 can fare at least equally well by letting $\beta_2 = 0$. In general, if we have nested models like M_1 and M_2 in this case, the more complex model with more parameters always has a better or at least an equally good likelihood, because it has more degrees of freedom for accommodating the data. As long as we compare likelihoods, we always end up favoring the more complex model.

Furthermore, there is no direct relationship between a model's likelihood and its predictive performance, which is our primary concern here. Likelihood, recall, is the probability of obtaining the *given* data under the assumption of a particular hypothesis/model, so it measures how well the model makes sense of what is observed. In contrast, prediction concerns what the model will tell us about data *yet to be observed*. A model that accommodates the past well is not necessarily the best guide for the future. The reader may sense here the specter of Humean skepticism, but there is a further glitch. Even if we assume a uniformity of nature, fitting a model precisely to the data is not always a good idea. This is because any dataset from a stochastic process necessarily contains sporadic noise—hence, a complex model that accommodates the given data well may also have been fit to this noise, which would compromise the model's ability to predict unobserved data. This is called *overfitting*. In order to avoid overfitting, we need a criterion other than the likelihood for evaluating a model's predictive ability.

4.2.3 Akaike Information Criterion

Akaike's approach to this problem is to turn our eyes from the model's likelihood to the *mean* likelihood (or more precisely, mean log likelihood). We are interested not in how well a model accommodates the actual data at hand, but rather in how well it predicts data yet to be observed. But how to measure this predictive ability? One reasonable idea is to consider the model's *average* predictive performance that *we would obtain if we were to keep using it to repeatedly predict similar datasets*. If we measure the performance of each prediction by the likelihood, this amounts to evaluating the model's mean likelihood.

As a specific example, consider making similar predictions over and over again using the model M_1 (see Figure 4.2). We begin by fitting the model M_1, i.e., calculating the MLE of its parameters, to the data of, say, 1000 college students in a particular year. Let us denote the resulting fitted model by $\hat{M}_1$. Since the parameters of $\hat{M}_1$ are fully specified, we can use it to predict new data. So we collect data for another 1000 students the following year and calculate the likelihood of this fitted model using this new data. Suppose we

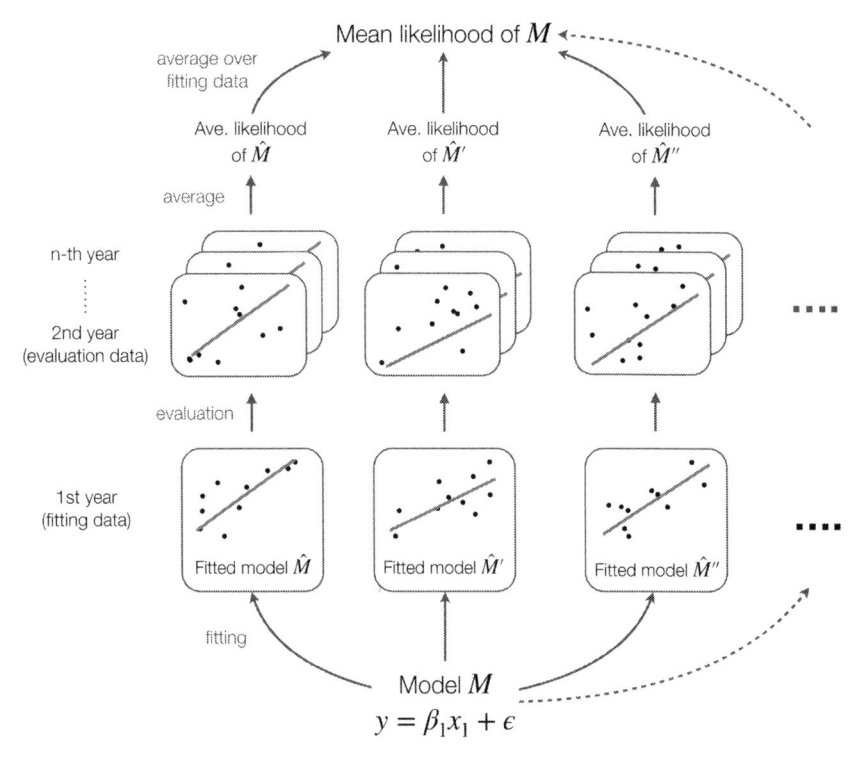

FIGURE 4.2 A (hypothetical) calculation of a model's mean likelihood. The predictive performance of a *fitted* model $\hat{M}$ can be evaluated by averaging its likelihood with respect to many datasets of the same nature. The mean likelihood of the model M is obtained by repeating this process over different initial datasets, using fitted models $\hat{M}', \hat{M}'', \ldots$ (note these have different regression slopes due to randomness in the fitting data). Since this calculation is infeasible in reality, AIC aims to estimate it from a single dataset.

repeat this procedure of calculating the likelihood the next year, the year after that, and so on indefinitely. By averaging the model's likelihoods over n years, or theoretically over infinitely many years, we will obtain the mean likelihood of the fitted model $\hat{M}_1$. This already seems like a lot of work, but we're not there yet: what we have calculated is the mean predictive performance of the fitted model $\hat{M}_1$, which was obtained by attuning the model M_1 to the particular dataset we happened to observe in the first year. Because this initial dataset clearly involves some randomness, there is no guarantee that this particular result correctly measures the predictive performance of M_1. We thus need to average out this initial variability too, by repeating the entire fitting procedure with many different initial datasets. This will eventually give us the mean likelihood of the model M_1. We may take this as a measure of the model's predictive ability, in the sense that it tells us how well the model is expected to make predictions on average, if it is used repeatedly for similar predictive tasks.

To derive the mean likelihood, we had to imagine making indefinitely many predictions with the same model and taking the expectation of the outcomes. Since such a maneuver cannot be done in practice, the mean likelihood of a model is not something one can measure directly from data. But like other parameters in general, we may *estimate* it from the data at hand. Akaike has shown that under certain assumptions,[7] an estimator of the mean log likelihood of a model with k parameters is given by

$$\log \ell(M) - k$$

This shows that two factors affect a model's mean predictive performance. One is the model's maximum log likelihood, $\log \ell(M)$, which expresses how well the model M accommodates the data at hand. As we noted previously, as the model becomes more complex, this term increases and contributes positively to its predictive performance. But since this value is calculated on the basis of a particular dataset, there is no guarantee that the same model will score an equally good likelihood when calculated with new data. In this sense, the actual likelihood overestimates the model's *mean* predictive performance, so it needs to be discounted. The correction comes from the other term k, which represents the number of parameters in the model. Complex models have many parameters, and thus larger k. The minus sign before k means that too large a k compromises predictive performance; this term thus imposes a penalty on the complexity of a model. As a result, a model's mean log likelihood is determined by the balance between its data-accommodating ability afforded by its complexity, and the penalty placed on this complexity. With our example models M_1 and M_2, the first factor favors M_2, since $\log \ell(M_1) \leq \log \ell(M_2)$. M_2, however, has $k = 4$ parameters whereas M_1 has just $k = 3$, so the second factor favors M_1. Therefore, the more complex model M_2 ultimately outstrips M_1 if the introduction of the extra parameter in M_2

increases its likelihood enough to compensate for the penalty placed on the extra parameter.

Although this discussion featured *nested* regression models, where one model (M_1) is a proper subset of the other (M_2), Akaike's framework can also effectively evaluate the relative predictive performance of non-nested models or different distribution families. Conventionally, the mean log likelihood times −2, namely

$$-2(\log \ell(M) - k),$$

which is now called the Akaike Information Criterion or AIC, is used to evaluate the predictive performance of models. We cannot go into the details here, but AIC gives, under certain assumptions, an unbiased estimator of the mean discrepancy (measured by the so-called Kullback–Leibler divergence) between the model's predictions and random samples taken from the true probability model (Konishi and Kitagawa 2008). We can thus expect that a model with a smaller AIC gives better predictions, in the sense that its predictions deviate less on average from actual sampling outcomes.

4.2.4 Philosophical Implications of AIC

The insight of the theory of AIC discussed in Section 4.2.3—namely, that too many parameters may impair a model's predictive performance—has a somewhat paradoxical implication. According to the framework of traditional inferential statistics that we saw in the previous chapters, a statistical inference proceeds by modeling the data-generating process in terms of a certain probabilistic kind (statistical model), whose details are then specified from the data to make predictions of unobserved samples (see Figure 1.2). This picture naturally leads one to expect that the more accurately the presupposed probabilistic kind approximates the actual data-generating process, the better its predictions become. The theory behind AIC, however, suggests that this expectation is not necessarily borne out. To see this, let us suppose that in reality our response variable Y is influenced by both explanatory variables X_1, X_2—that is, that the model M_2 (Equation 4.4) gives a true and complete picture of the underlying probability model. We further assume, however, that the influence from X_2 is much smaller than that of X_1, so that $\beta_1 \gg \beta_2 \approx 0$. In this case, adding the factor X_2 to the model as in Equation (4.4) will not boost the log likelihood. If the increase is less than 1, the model M_1 (Equation 4.3) with only one explanatory variable X_1 may have a smaller AIC score, and thus be judged to make better predictions (in the sense described earlier). This is so even if (by assumption) the model M_1 overlooks the actual factor X_2, and is hence more distant than M_2 from the actual data-generating process. AIC, therefore, indicates the possibility that a "true" statistical model that faithfully describes the underlying probability model may nevertheless fare worse in prediction than a model that "distorts" reality by omitting some factors.

The conclusion that a model removed from the truth can give better predictions may sound paradoxical to some. But far from being a paradox, this is actually a general feature common to any kind of scientific reasoning based on models or natural kinds. For one thing, proper idealizations and simplifications are part and parcel of all scientific investigation (Cartwright 1983). In effect, to classify worldly things into discrete natural kinds already involves an abstraction of individual niceties, and in that sense is a distortion of reality. You and I have different physical compositions, and even the cells constituting my skin differ slightly from each other. To classify these different things in terms of natural kinds such as "*homo sapiens*" or "epidermal cells" is to ignore their particularities and distort their details. But this kind of abstracting away is precisely what makes inductive inferences possible; for example, "what is toxic to me will also be so to you," or "an ointment that worked with an insect bite here will also work with one there." If we refuse to admit this kind of rough categorization and insist on counting every single human as a distinct being, we will no longer be able to learn anything from the experiences of others. To perform inductive reasoning, therefore, we need to identify things at a certain level of granularity and ignore all niceties and peculiarities below it. The same applies to statistical reasoning with probabilistic kinds. The same data-generating process can be described by different models with different numbers of parameters. The question of which probabilistic kind to use is comparable to the question of at which granularity we should describe a human being, where the possible options would include "living creature," "animal," "mammal," "*homo sapiens*," "middle-aged men," and so on. And just as an unnecessarily scrupulous natural kind does not contribute to inductive inferences, a probabilistic kind that is too detailed will not give us effective predictions. AIC makes this explicit in terms of the mean log likelihood and informs us of the probabilistic kind of the right granularity through the estimation of its long-term predictive performance.

This invites us to rethink the statistical ontology introduced in Chapter 1. As we noted there, probabilistic kinds or natural kinds in general provide basic ontological units with which scientists carve nature (Section 1.2.4). Such natural kinds often form a hierarchical structure. For instance, a human being, an example of an ecological natural kind, consists of cells and other physiological kinds, which in turn are composed of chemical kinds such as molecules and atoms, which are further composed of physical kinds such as protons, electrons, and so forth. Lower-level kinds compose higher-level ones and allow for finer-grained descriptions of reality. If all entities existing in nature are composed of physical elements, then everything should be describable solely in terms of physical kinds. Why, then, do we still care about naive and coarse-grained natural kinds like "trees" or "birds," rather than regard them simply as clouds of particles condensed in different ways? The reason is that higher-level natural kinds, though they may be vague and imprecise descriptions of the world, play an indispensable role in explanation and prediction. By identifying a bird that built a nest on my neighbor's tree as a swallow, I can predict its departure in autumn. I certainly could not make a similar prediction at the atomic level. Virtually all our everyday inferences are made possible in this way, i.e., by demarcating a part of the world and subsuming it under a certain category. Daniel Dennett (1991) called these categories that we carve out from nature *real patterns*. "Swallow" and "noble metal," along with other natural kinds, are typical examples of real patterns that help us in our predictive and explanatory endeavors in relevant contexts. When we look at them closely, they may be very rough generalizations splattered with noises and exceptions. Be that as it may, they are real insofar as they help us predict the future, and in this sense they have every right to be treated as *bona fide* entities.

Going back to statistics, we observe that some probabilistic kinds form a hierarchical structure just like conventional natural kinds. One and the same relationship between two variables could be modeled using a linear regression model, or with a polynomial model that accommodates any smooth curve; the latter can describe the relationship with much better resolution and precision. If our goal is to obtain a faithful reproduction of the uniformity of nature, then we should prefer complex models with a higher degree of freedom, since they give a picture closer to the truth. If, however, our goal is prediction, a model that is too fine-grained may fare worse than a coarse-grained one due to random noise inherent in the sampling and estimation processes. Faced with this situation, AIC aims to identify the model with the right granularity, or in Dennett's parlance, to carve out a real pattern from the data. Note the two meanings of "real" here. In one of its senses, the word reflects the idea that a model that approximates the data-generating process (i.e., probability model) well is real, while in the other sense it captures the Dennettian idea that patterns that contribute to predicting unobserved instances should be regarded as real.

Distinguishing these two ontological attitudes helps us understand what AIC is and what it is not. It is sometimes claimed that the purpose of AIC is not to choose the true model (Kasuya 2015), while at other times it is said to measure the (expected) distance of a model from the true distribution (Leeow 1992; Ponciano and Taper 2019). These apparently conflicting ideas nevertheless make a sense if we consider the dual meaning of "reality" mentioned previously. The word "true" in the first claim refers to the faithful reconstruction of the probability model that generated the given observations, and that is certainly not the goal of AIC. In contrast, the "distance" in the second claim refers to the discrepancy between *predictions* from a given statistical model and future samples to be taken from the true distribution; in other words, it refers to the extent to which the model under consideration deviates from the pattern that we recognize at different times and places (Figure 4.3). AIC thus evaluates the reality of statistical models in the sense of real patterns, and by doing so it brings an alternative meaning of "reality" into statistics. One might even argue that this alternative ontology is more in line with the original aim of statistical inference. In Chapter 1, we characterized the probability model as a fundamental

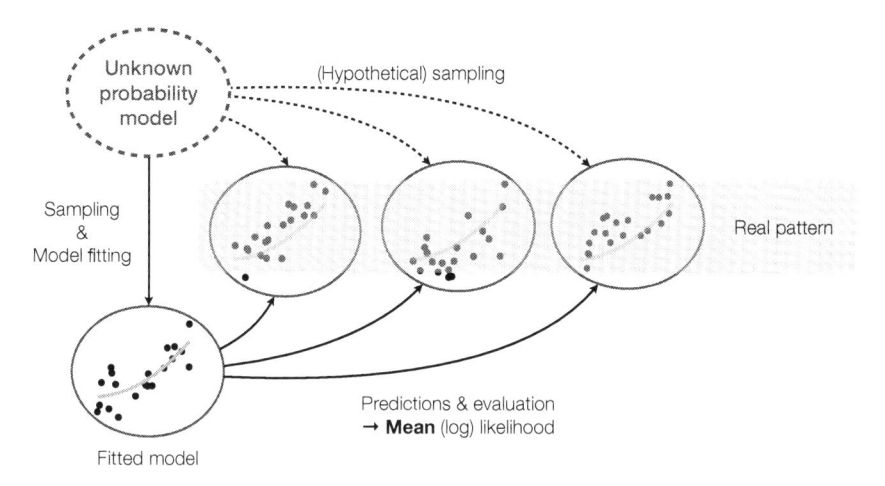

FIGURE 4.3 AIC evaluates models based on data obtained from an unknown probability model (the left vertical arrow). The goal is not to give a faithful description of the probability model, but rather to choose a model that accurately predicts, on average, future data (right) to be obtained from similar sampling processes. This is tantamount to identifying a real and stable pattern that will appear over different datasets.

ontological assumption of inferential statistics. But why did we need to introduce an additional entity beyond the data? That was because inductive inference is impossible without this assumption of the uniformity of nature. This suggests that from the very first motivation for their introduction, the conceptual function of "entities" in statistics is not descriptive (i.e., to describe the data-generating process as it is), but rather instrumental (to make successful predictions). Given this purpose for introducing a new entity, it is only natural that an entity's contribution to prediction should be a major criterion when deciding to carve it into a probabilistic kind of a certain form. AIC gives substance to this idea by providing a method for evaluating the reality of probabilistic kinds *qua* real patterns.

Behind this discussion lie deeper questions regarding the nature and goal of statistics and of science in general. A popular image of science is that it aims to uncover the world as it really is in its finest details. On this view, natural kinds, being the basic conceptual units of science, must capture what the elements composing this world are really like. That is, the ideal scientific ontology must be able to reproduce the actual world down to the most minuscule details. It is by grasping the structure of the world in this way that science is able to explain worldly phenomena.

While this may indeed be a plausible picture, it is not the only view. Another perspective on science takes its major goal to be the elucidation of not so much how the world actually *is* as how it will *become*—that is, to make successful predictions (van Fraassen 1980). Francis Bacon's famous dictum, "*scientia potentia est*," refers to this predictive power of science. If we are to follow this Baconian line of thought, the aim of scientific ontology should be not so much the making of a faithful replica of the world, as the identification of real patterns through appropriate abstractions and simplifications. This leads us to a kind of *pragmatism* (Sober 2008). William James, one of the early founders of pragmatism, argued that we should replace the traditional conception of truth as a correspondence of our ideas with the external world with a new theory of truth, according to which the truth is nothing but those beliefs that serve us well (James 1907). On this view, the claim that such and such things exist is judged true if and only if such a belief contributes to a particular purpose, which in our context is prediction. It is not that true ideas guide our reasoning because they are true; rather, those ideas that facilitate inductive inferences are acknowledged to exist as natural or probabilistic kinds. Pragmatic scientific ontology thus inverts the relationship between existence and cognition.

What is worth noting here is that what counts as useful depends on the context. In the present context of prediction, the model that AIC chooses to recommend depends, among other things, on the size of data we can afford in making predictions (Figure 4.4). As the data size becomes larger, the absolute value of a model's log likelihood increases. This means the penalty on the number of parameters will weigh less, swinging the balance in favor of a more complex model. By the reverse logic, small datasets will tend to favor parsimonious models. Which probabilistic kinds are picked out as real patterns, therefore, depends not only on the objective features of the world, but also on a practical factor—the size of available data. Due to this feature, AIC has sometimes been criticized as lacking *statistical consistency*. An estimator is said to be consistent if it converges to the true value as the data size approaches infinity. For instance, Bayesian inferences are consistent in that the posterior probabilities converge to the true distribution as the sample size grows indefinitely (see Section 2.3.3), and in the same vein the Bayesian model-selection criterion (so-called BIC) asymptotically selects the true model. This is in contrast with AIC, which, for the reason stated earlier, does not necessarily choose the model that faithfully captures the data-generating process, even with an infinitely large dataset (Sober 2008). However, there is nothing problematic about this given the purpose of AIC, which is to identify real patterns that contribute to prediction. Since any prediction must be made on the basis of a limited data source, the pattern that counts as "real" should depend on the user, or more precisely, on the amount of data available to the user, just as the patterns of odor that dogs likely recognize as real may well differ from those we humans are able to identify. From the canine perspective, the patterns we are able to sniff out with our poor olfactory sense might well be miserably coarse-grained. But as long as it helps us predict today's supper, the smell of, say, clam chowder or apple pie is a *bona fide* real pattern. If the odorous patterns to be counted as real depend on the number of olfactory cells that a cognizer (dogs or humans) possesses, it is only natural that the probabilistic

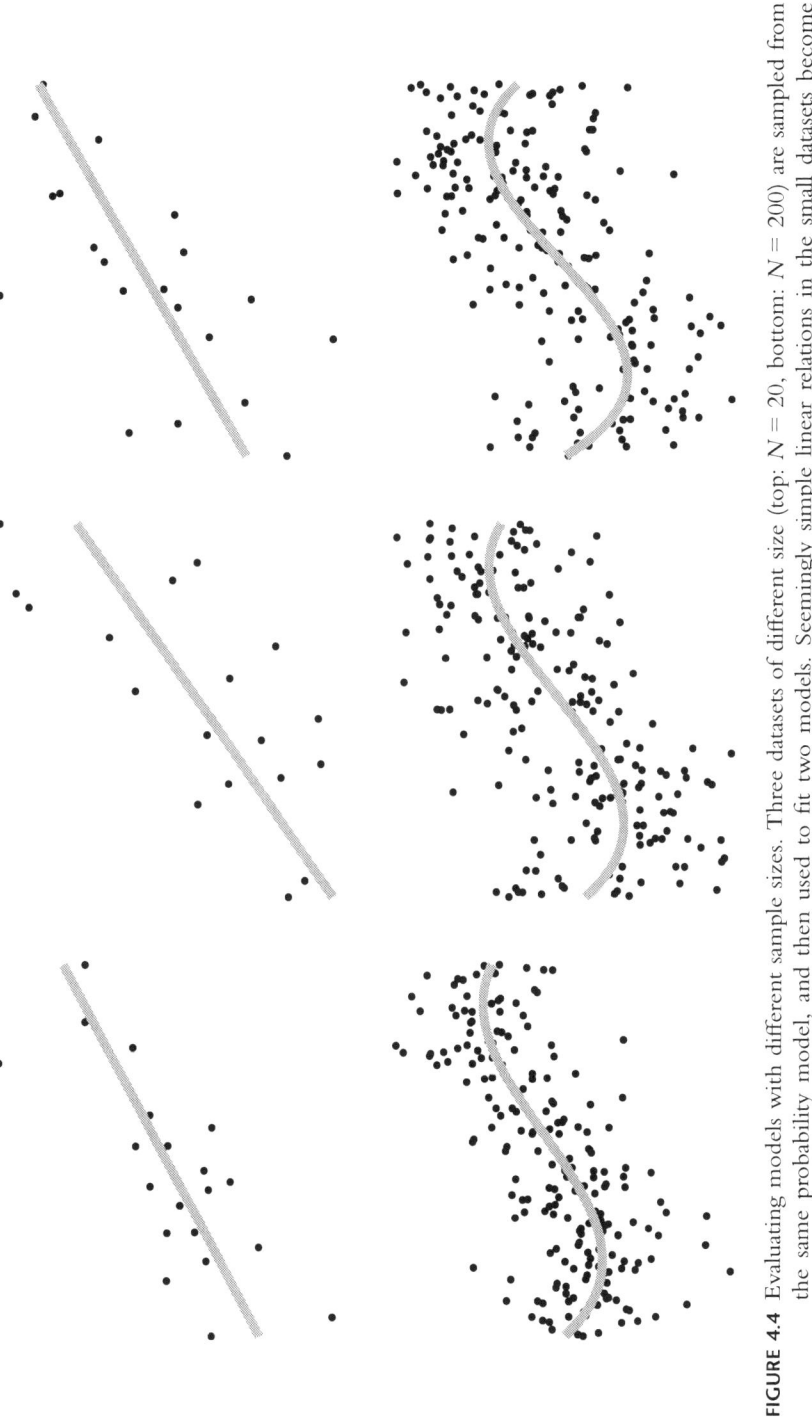

FIGURE 4.4 Evaluating models with different sample sizes. Three datasets of different size (top: $N = 20$, bottom: $N = 200$) are sampled from the same probability model, and then used to fit two models. Seemingly simple linear relations in the small datasets become more like cubic curves in the large ones. AIC favors the linear model in the former and the cubic model in the latter, suggesting that the pattern that counts as "real" varies according to the size of data.

kinds to be counted as real likewise depend on the amount of data the cognizer has. In this spirit, AIC carves the world in line with its pragmatic goal of predicting the future based on limited resources. This kind of pragmatic ontology also implies the possibility that different real patterns emerge for different cognizers. We will return to this ontological relativity after we review deep learning, which in a sense takes the opposite approach to the problem of prediction.

4.3 Deep Learning

Prediction is a central topic not only in model selection, but also in the rapidly developing field of machine learning, the most successful among which is *deep learning*. In this section we review its basic machinery and consider how it approaches the problem of prediction.

4.3.1 The Structure of Deep Neural Networks

Although deep learning has come to be applied to various kinds of problems, its central task has been the problem of prediction. Image classification, automated medical diagnosis, and speech transcription are examples of predictive tasks, where the goal is to return the most appropriate value y of the response variable as output given some values x of the explanatory variables as input. In this sense, a deep learning model adapted to this sort of problem is a kind of regression model. But in contrast to the conventional parametric statistical models, deep learning models usually have an extremely complex structure with an enormous number of parameters, and in this respect they go against AIC's spirit of "less is more." On the other hand, its fitting method incorporates an idea similar to the one we saw in Akaike's theory. With this (dis)similarity in mind, let us briefly review the structure of deep learning models and the methods used to train them.

The standard model of deep learning is a *deep neural network* (Figure 4.5). Each node (neuron) of a neural net represents a random variable. These nodes are arranged in layers, which are piled up to form a multilayered network. The model successively computes values of the variables/nodes from the left layers in the figure to the right, until the rightmost layer gives the final output. First, the data are fed into the leftmost input layer $(X_1, X_2, \ldots, X_N)$; these correspond to the explanatory variables of a regression model. The only difference is the scale—while the number of inputs or explanatory variables in most traditional regression models ranges from several to at most a few dozen, in deep learning models the number far exceeds thousands or even millions. Even the simplest model that recognizes an object in a 256 × 256-pixel monochrome picture has 65,536 input variables, each of which quantifies the brightness of a pixel at a particular location. Commonly used images, videos, and voice data require much

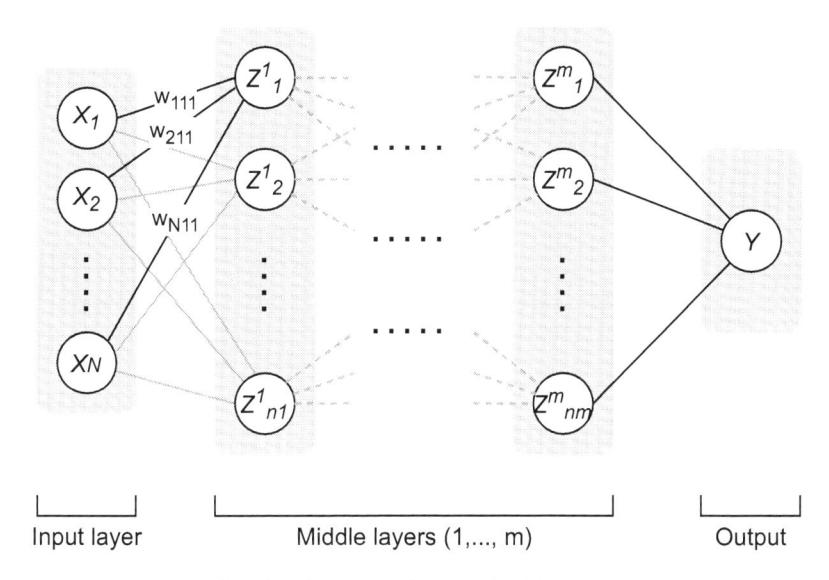

FIGURE 4.5 An example of a deep neural network. The input layer $\boldsymbol{X}$ is projected onto the middle layers $\boldsymbol{Z}^1, \ldots, \boldsymbol{Z}^m$, which finally lead to the output Y. Each projection constitutes a regression model. The nature of the projections and structure of the layers differ from model to model. A model whose nodes in consecutive layers are fully connected as in this example is called a multilayer perceptron.

larger inputs. Next, the variables in the input layer are projected onto the adjacent middle layer $\left(Z_1^1, Z_2^1, \ldots, Z_{n^1}^1\right)$. (Here the superscripts denote the fact that these variables belong to the first middle layer, while the subscripts label the n^1 nodes/variables constituting this layer.) The values of these middle-layer variables are determined by the values relayed from the input layer. In fact, each of these determination processes constitutes a regression model like those we saw in the previous section. For example, the value of the topmost variable Z_1^1 can be written as

$$z_1^1 = f\left(w_{011} + w_{111}x_1 + w_{211}x_2 + \cdots + w_{N11}x_N\right) \quad (4.5)$$

Note that what is inside the parentheses is nothing but the now-familiar linear regression model. The intercept w_{011} is the baseline, while the ith regression coefficient w_{i11} represents the weight of input x_i. The value of Z_1^1 is obtained by applying an activation function f to the weighted sum of the inputs. There are several kinds of activation functions, but the most commonly used is the rectified linear function/unit (ReLU):

$$f(u) = \max(u, 0),$$

which simply returns zero if the inside of the parentheses in Equation (4.5) is smaller than zero, and returns the input value itself if it is zero or larger. One can think this as a neuron that "fires" if its input exceeds the threshold of zero, but stays inactive otherwise. Repeating this calculation for each variable in $Z_1^1, \ldots, Z_{n^1}^1$ determines the values of the first middle layer.

A deep neural network is constructed by piling up such layers over and over. Each middle layer is calculated in a similar manner: the nodes in the $j + 1$-th layer are determined by a regression model like Equation (4.5) that takes inputs from the nodes in the *j*th layer. In the final step, the terminal middle layer is connected to the output *y* via another regression model. This regression model that gives the final output comes in a variety of forms, depending on the purpose of the model. When we want to make a prediction for a continuous variable, for instance, we might use a simple linear regression, whereas for a classification into discrete categories, one often uses logistic or softmax functions that return certain labels according to thresholds.

In this way, a deep neural net is constructed by stacking layers of nodes connected via regression models. The model as a whole takes the explanatory variables X as input, processes them successively through the middle layers, and finally yields outputs in the response variable *Y*. This means that a deep neural network is itself a gigantic regression model. Indeed, the whole model can be written down in a compact form as

$$y = g(x; w) \tag{4.6}$$

This shows that the behavior of a deep neural net *g* is specified by the set of parameters w, and that its output *y* is uniquely determined given the input x. Hence, in the parlance of this book, a deep neural net is simply another probabilistic kind, albeit an enormous and labyrinthine one. While the size of neural nets varies depending on their purpose, the standard models in the deep learning literature (at the time of writing) have tens or hundreds of middle layers, with parameters ranging in the billions. Thus, despite its simple guise, Equation (4.6) may contain billions of parameters w, and the function *g* usually has such a complex form that it is impossible to write it down as an explicit formula.

4.3.2 Training Neural Networks

Constructing a deep neural network in this way is only the first step. The next important step is to train the model, i.e., to adjust its parameters by fitting the model to concrete data (called "training data"). At the beginning of this chapter, we reviewed the method of maximum likelihood as a paradigmatic approach to model fitting. Recall that the method regards a model's likelihood as a function of its parameters and searches their maximum points (that is, the combination of parameters that maximizes the probability of the data) by a step-by-step "climbing" of the likelihood function. The same approach is taken in the training of deep learning models, but it is customary to use the negative of the log likelihood and train the model by *minimizing* it by "descending" into a valley. Flipping the sign in this way does not at all change the essence of the maximum likelihood method. The target negative log likelihood function is called the *error function* or loss function since it expresses the degree in which the model *fails* to fit the data. Other common measures of misfitting include the aforementioned method of least squares. In this case, the goal is to find the parameters that minimize $\Sigma_i^2 (y_i - \hat{y}_i)$, the square of the difference between the model's output $\hat{y}_i$ and the actual value y_i summed over every sample i in the training data.

The basic strategy for fitting deep learning models, therefore, does not differ from those used in traditional statistical models. The complexity and sheer magnitude of deep neural nets, however, pose unique difficulties. First, the immense number of parameters means that the training by "valley descent" must take place in a space of enormously large dimension, of the order of millions and billions. We saw previously that the numerical approach for the method of maximum likelihood sets off from a particular combination of parameters taken as a starting point, differentiates the log likelihood or loss function to calculate the slope (gradient) at that point, and then moves along the calculated slope—repeating this process should gradually increase the likelihood or decrease the loss. To carry out this process over the many layers of parameters, deep learning adopts the *backpropagation method*. This technique starts by calculating the derivatives of the loss function with respect to the parameters situated right next to the output layer, and then propagates the results toward the input. We will not go into the details here, but such a cascade-like calculation procedure allows us to determine the slope of the loss function at any point, even for a model with billions of parameters. So in theory it works beautifully, but in actual practice we are faced with a technical issue: the backpropagation method amounts to multiplying the derivatives of parameters many times through multiple layers from output to input, and in this process the values may vanish or diverge to infinity. This *vanishing gradient problem* especially vexed early attempts to train multilayered neural models (e.g. Kelleher 2019, ch. 3).

Determining the slope of a loss function is not the whole story. As mentioned before, whether the step-by-step valley descent can bring us straight to the optimal point critically depends on the shape of the valley; that is, the nature of the loss function. In this regard, it is very unlikely that the loss function of a complex deep learning model has a simple form like in Figure 4.1; instead, its landscape is expected to be rugged, with precipitous mountains and deep valleys, making a straightforward hike to the peak hardly possible. Hence, wherever one starts, one can hardly hope to arrive at the global optimum and must be content with a locally "good enough" solution. This results in underfitting, but there is also the opposite risk of overfitting. As we saw in the previous section, increasing the number of parameters allows the model to better accommodate data but may also make it sensitive to noises peculiar to the training data, thereby impairing its performance in predicting new data (called its "generalization capability"). The enormous number of parameters in deep neural networks make them highly flexible and capable of approximating any function to the desired accuracy. But due exactly to this *universal approximation property*, deep learning models are prone to overfit to particular datasets and lose their generalization capability. These issues posed serious obstacles to early research on deep neural networks.

The remarkable advances in deep learning today owe to various breakthroughs with regard to these problems. There are roughly three classes of approaches and techniques for effectively training huge neural nets without falling prey to the vanishing gradient problem or overfitting:

1. *Designing the network architecture and its components in such a way that large-scale models can be effectively trained.* The model in the earlier illustration is fully connected, with each node in one layer affecting every node in the next layer. This, however, is not necessary. For instance, one may reduce the number of connections and parameters by allowing the nodes in a middle layer to take inputs only from a part of the previous layer. A paradigmatic example is convolutional neural networks (CNN), which demonstrate high performance in image recognition and can be trained effectively even when the model has many layers. The choice of activation function (Equation 4.5) is also important. For example, the vanishing gradient problem was significantly improved by the introduction of the aforementioned rectified linear function, which replaced the sigmoid or logistic function that had been previously used.
2. *Partial or piecemeal training of models through pre-learning, dropout, and batch normalization.* A major difficulty in training deep neural nets stems from the fact that, due to the immense number of parameters, adjusting parameters in one part of the model tends to interfere with the adjustment of parameters in other parts. This can be partly prevented by decomposing the model into manageable portions and training them part by part. For example, training each layer of a network before constructing and training the entire model is known to mitigate the vanishing gradient and overfitting problems. The dropout technique, on the other hand, randomly picks out some nodes in the neural net to be adjusted while ignoring the rest; by repeating this process, it trains the entire model in a piecemeal fashion.
3. *Modifying the loss function (i.e., the optimization target) to reduce the degrees of freedom.* A typical method is to put a cap on the magnitude of parameters when one tries to optimize the likelihood or mean squared error. One may, for instance, minimize the original loss function *plus* the sum of squared parameters $\sum_i w_i^2$, so that models with smaller parameter values are favored. Like AIC in the previous section, this constrains a model's degrees of freedom by

putting a penalty on parameters (though in this case, what is penalized is not the parameter count but their magnitude). This kind of technique is called *regularization*.

The methods listed here are not exclusive (or exhaustive, needless to say), but are generally used in combination. These techniques now allow researchers to train even huge deep learning models with millions or billions of parameters and achieve high generalization performance, without falling prey to overfitting.

4.4 Philosophical Implications of Deep Learning

4.4.1 Statistics as Pragmatist Epistemology

The recent progress of deep learning and the accompanying boom in artificial intelligence (AI) are rapidly changing our everyday life and even the fabric of society. Their success prompts a sort of paradigm shift from the traditional Bayesian and classical conceptions of statistics. The shift, in a word, is one from truth to predictive performance. As we have seen so far, the major goal of traditional inferential statistics is to identify the uniformity of nature behind the data, or an ideal model of it (aka a probabilistic kind). Prediction is considered a subtask of this identification problem and is approached indirectly via an inference to the probabilistic kind that mediates between the observed and unobserved samples (see Figure 1.2). In this sense, one can say that the first and foremost objective of traditional statistics, both Bayes and classical, is to build a true and faithful picture of the world.

In contrast, obtaining a true model is not the primary goal in model selection or deep learning. Instead of taking models as pictures of reality, AIC and other model selection criteria regard models as tools and evaluate their instrumental value in terms of their predictive performance under given data (Section 4.2.4). The same applies to deep learning. True, the universal property of deep learning models allows them in principle to represent any complex data-generating process in the finest detail. There is no guarantee, however, that models can actually be trained to attain the true distribution; indeed, given the vast dimensions of the search space and the complexity of the loss functions, an exact match is highly unlikely (Goodfellow, Bengio, and Courville 2016, sec. 8.2.2). This, however, poses no practical problem provided that the models demonstrate satisfactory performance in predictive tasks. Again, the focus here is on a model's utility rather than veracity.

What is the epistemological implication of this shift from truth to prediction in contemporary statistics? In other words, what kind of epistemology do the theories of model selection and deep learning involve? One natural and sensible approach to this question would be to take a cue from *pragmatist epistemology*, which was proposed as an alternative to traditional epistemology toward the latter half of the 20th century. The traditional epistemology we have discussed so far has tacitly assumed the acquisition of truth, understood as a correspondence between our thought and the world, as its primary objective. Both internalists and externalists take knowledge as (at least) a true belief and understand justification as a means for arriving, in some way or another, at the truth. Underneath this epistemological agenda is the conception of cognition as a "mirror of nature" which supposedly reflects the world as it is (Rorty 1979). According to this view, the goal of our cognition is to produce a faithful and accurate image of the external world within the mind, like a reflection in a clear, stainless mirror. Though it may appear natural, Rorty points out that this view itself is an unverifiable metaphysics, built upon the trite dogma of the distinction between subject and object, or between minds and the world. Accepting such an epistemological framework inevitably leads us to the skeptical aporia of how we can ever know whether our cognition faithfully reflects the external world or not. Rorty claimed that to shake off this aporia, we need a different, pragmatist, epistemology. As we have seen, pragmatists have us understand beliefs and concepts in terms of their potential consequences on our behavior and practice. Those who believe that glass is fragile and those who believe that glass never breaks will likely behave differently in everyday life and get different payoffs. The latter people may lean their entire body weight on a window or try to drive a nail with a piece of glass, and as a result may get injured or even lose their life. Such a belief is obviously not pragmatic. On the other hand, those who believe that iron is harder and less fragile than glass would be better able to attain their end by using a hammer. In this way, beliefs have instrumental value, some being more pragmatic than others. Pragmatists claim that our beliefs should be evaluated not in terms of their veracity—whatever that may be—but rather in terms of this kind of utility. In other words, truth is not an intrinsic or primary value of a belief; the "goodness" of a belief should be determined not by its closeness to the truth, but by whether or not it leads us to desirable consequences in our daily lives or in scientific practice.

If, as pragmatists say, truth has no intrinsic value, then the whole concept and aim of epistemology must undergo a complete overhaul. It has been customary in traditional epistemology to evaluate epistemic systems in terms of their truth-conduciveness, i.e., how well they allow us to obtain true beliefs. But if the value of beliefs is to be measured not by their veracity but by other pragmatic criteria, epistemic systems too must be evaluated in terms of their contributions to those diverse goals. This leads Stephen Stich, a prominent pragmatist epistemologist, to claim that "Cognitive processes . . . should not be thought of primarily as devices for generating truths," but "as something akin to tools or technologies or practices that can be used more or less successfully in achieving a variety of goals" (Stich 1990, p. 131). Indeed, we use our cognitive machinery for various purposes in our everyday life and work, such as driving a car through busy traffic, cooking a fancy dinner, or gaining territory in a *Go* game. These tasks all demand sophisticated cognitive processes, but of different natures and functionalities; so it would be rather unfitting to assess them using the same measure of "truth." Each task has its own goals and values, and therefore its own criteria for evaluating the cognitive means used. There should be no overarching gold standard applicable to all kinds of cognitive processes; rather, the goodness of a cognitive process must be judged in terms of how much it contributes to a given concrete task, like driving safely, preparing tasty meals, or winning a *Go* game.

This kind of pragmatist epistemology appears to fit well with the trend of contemporary statistics exemplified by machine learning. To be sure, model selection and deep learning do not dispense with the notion of a true distribution. The "truth" there, however, is no longer considered to be an ultimate goal of inquiry, but rather as an instrumental assumption used to derive effective predictive methods and to assess their performance (Konishi and Kitagawa 2008, p. 3). The goal of model-building on this view is not so much to obtain a faithful picture of the data-generating process as to make successful predictions and classifications. This is the crux of Box's dictum that we saw in Chapter 1: "all models are wrong, some are useful." "So," he continues, "the question you need to ask is not 'Is the model true?' (it never is) but 'Is the model good enough for this particular application?'" (Box, Luceño, and Paniagua-Quinones 2009, p. 61). This passage embodies the pragmatist spirit that the value of a model consists not in its truth, but in its utility in solving concrete tasks. And to further back up Stich's point, the reasons one uses statistical models today are not limited to the pursuit of truth. As is well known, deep learning models are remarkably successful in a variety of applications, including automated driving, machine translation, playing *Go* and video games, drawing and painting, music composition and performance, and so on. Accomplishing such complex tasks surely requires highly sophisticated cognitive faculties and skills. Such cognitive capabilities may be comparable more to the *techne* we find in skilled craft-workers, athletes, and artists, rather than to the *episteme* involved in uncovering the truth or laws of nature. Aristotle said that *techne*, or art, is evaluated not in light of necessary and universal truth, but by its contribution to a given end (*Nicomachean Ethics* 1139a20–1140a20). Likewise, deep learning models are assessed with respect to the loss function determined by the technical problem they are supposed to solve. But from the pragmatist perspective, there is no essential distinction between technical applications and "pure" intellectual activities—they are both *bona fide* forms of inductive inference, evaluated by the same, pragmatic, standard.

4.4.2 The Epistemic Virtue of Machines

With its emphasis on the pragmatic role of the human as well as machine intellect, irreducible to the search for truth, pragmatism seems capable of doing justice to one significant aspect of modern statistics that does not fully square with traditional epistemology. But even if we acknowledge the pragmatist spirit at the root of the success and spread of modern machine learning, it is premature to conclude that contemporary statistics has completely lost touch with the quest for truth. Just as traditional statistics had evolved through both industrial applications and scientific inquiry (see Section 3.3.2), deep learning is also beginning to be applied to a diverse range of scientific research (Bianchini, Müller, and Pelletier 2020; Tanaka, Tomiya, and Hashimoto 2021). Machine learning is expected to help or even replace human reasoning in handling ever-increasing data and complex hypotheses in a variety of scientific disciplines, including physics, chemistry, biology, economics, sociology, and even the humanities. Certainly, the primary goal of such scientific applications is to elucidate the phenomena and principles in the domain of inquiry, that is, to attain the truth. Iten et al. (2020), for instance, report that a deep learning model has by itself rediscovered physically meaningful parameters and laws just from observed data, without any input of prior knowledge. Besides such exploratory use, machine learning is expected to gradually take over, at least to some extent, the role played by traditional Bayesian and classical statistics in the context of scientific justification. For instance, one may be able to use a machine learning model to judge among different scientific hypotheses vis-à-vis extremely complex big data. These applications will not only boost research activities but may also change the practice and even the very concept of scientific investigation.

This new direction in scientific research, however, confronts us again with the epistemological problem of justification that we have discussed throughout this book. Machine learning models may well analyze big data impenetrable by human understanding and give solutions to complex problems. But can we count such answers provided by computers as a piece of "knowledge"? Suppose that in the near future, some machine learning model discovers new physical laws. Can we then say, with this discovery, that we have extended our understanding and knowledge about the physical world, just as we did with the discoveries of Newton and Einstein? Or, in more general terms, are discoveries and conclusions given by deep learning models epistemologically justified, and if so, in what sense?

The reliabilist idea discussed in Chapter 3 seems to provide a natural first step to tackling this question. According to reliabilists, beliefs are justified if they are generated from a reliable epistemic process. Now, deep learning models do seem to be a highly reliable process, perhaps more so than our own, in helping us make appropriate decisions under complex situations. Granted, these mechanical "processes" differ from our innate cognitive functions, like vision and memory, in that they are external to us. That, however, does not matter to reliabilism; for one thing, a large part of our everyday cognitive decisions also depend on countless external processes like telescopes and computers. Long before the advent of machine learning technology, we humans have incorporated external tools to enhance our own cognitive processes. Moreover, seen as epistemic processes, there may not be any intrinsic difference between "internal" or innate cognitive organs and "external" deep learning models after all. What we usually call "innate" epistemic processes such as our sense organs, memory, and reasoning capabilities are all products of evolution and learning—that is, traits that have been formed through the optimization processes of natural selection and repeated trial-and-error.[8] Likewise, the cognitive functionality of deep learning models is built by a similar optimization process, as we have seen. The similarity is particularly striking in the method called *generative adversarial networks* (GANs), which improves the performance of models by subjecting them to mutual competition, just like an "arms race" in adaptive evolution. Innate organs "internal" to ourselves and "external" deep learning models, then, are in effect on a par, insofar as they are both cognitive processes produced through optimization processes. In this light, reliabilism seems well able to accommodate deep learning models into its justification process as part of our extended cognitive functions.

If so, the conclusions of reliable deep learning models may well be said to be justified in the externalist sense. That, however, is not the end of the story. Recall our discussion in Chapter 3, where we saw that inferences in classical statistics are reliabilist processes of justification. The key there was the theoretical warrant, provided by the mathematical theory of classical statistics, that backs up the reliability of tests and other statistical inferences. That is, there is a firm theoretical basis on which one can make a principled assessment of the reliability of a given test, using the concrete indices of confidence coefficient and power. Deep learning, on the other hand, still lacks such a foundational theory that would compute the reliability of models in a unified and *a priori* fashion. The performance of most deep learning models is instead evaluated *a posteriori*, by gauging how well they are able to perform tasks using standard datasets like MNIST or ImageNet—one might say, for example, that a new model has classified about 95% of the images in a dataset correctly. Arguably, properly performed benchmark tests may be taken as empirical proof of reliability, which would to some extent justify conclusions from models that achieve or exceed the SoTA (state of the art). These "proofs," however, must be obtained on a case-by-case basis and interpreted relative to a particular model and test set. In the absence of a unified theoretical standard, therefore, the reliability of machine learning models is attributed to each individual model in a rather makeshift manner. This leads to the "branding" of reliable models, because what is gauged by an individual benchmark test is the very model being tested and its characteristic properties. This is evidenced by the fact that deep learning researchers often give their models colorful names such as (Goog)LeNet, Alpha Go, BERT, and GPT-3 and discuss their performance and reliability under their name or brand. Those models have been acknowledged for their reliability and high performance in the standard benchmark tests and numerous empirical applications. What is suggested by all this is that the reliability of deep learning models is understood, at least in the current situation, as a property or skill peculiar to individual models.

What should we make of this idea of reliability as something inhering in individual models, and the concept of justification based on it? One clue for thinking about this question may be found in a recent trend in philosophical epistemology called *virtue epistemology* (Zagzebski 1996; Sosa 2007). Virtue epistemology locates the ground of justification in the nature or personality of epistemic agents, or their epistemic virtue. Let us illustrate the idea with a simple example in ethics, where this kind of view originated under the name of "virtue ethics." Ethicists have long disputed over what makes one's act morally "good." Imagine two people who donate to the needy; one donates reluctantly from a sense of duty, while the other donates from an unselfish, philanthropic spirit. Which one is morally better? From a utilitarian perspective, as long as their donations bring about the same benefit, there is no moral difference between the two, regardless of their motives. In contrast, virtue ethicists give higher praise to the latter, since that person's deed comes from his or her virtuous mind. According to virtue ethics, an action or behavior is morally good when it is a manifestation of an agent's virtue, such as benevolence, graciousness, conscience, and so on.

Returning to epistemology, virtue epistemologists apply this idea and consider a belief to be justified when it is a manifestation of the believer's epistemic virtue. What counts as epistemic virtue differs among theorists, but most often cited are perceptual acuteness, memory, inferential capability, curiosity, fairness, modesty, and so forth. Although the word "virtue" may sound pompous and somewhat preachy to some, there's actually no implication of the sort: the "virtue" in virtue epistemology is no more than the ability or excellence of a person in recognizing and understanding things. Virtue epistemologists grant that a belief is justified when it is obtained by such abilities possessed by an epistemic agent. Indeed, the way we appreciate an expert's advice more than a layperson's opinion in our everyday lives as well as in technical issues seems to reflect the fact that we implicitly adopt such a justificatory concept. Imagine that I, a complete layman about birds, go on a hike with my ornithologist friend and see a bird at a riverside. Suppose I judge that it is a kingfisher (since this very name suggests to me that they likely live near a river or sea). My ornithologist friend makes the same judgment, but hers is based on her acute eyes as a skilled birdwatcher as well as her expert knowledge about the bird's ecology, environment, and climate factors. You might then think that my belief about the kingfisher is not justified, or at least less so than hers. For one thing, while my judgment is just a sheer guess, my friend's is grounded on her expertise or epistemic virtue as an ornithologist. It is our general tendency to accept judgments of attentive, smart, knowledgeable people as more justified. Virtue epistemology features such inherent and personal characteristics that we implicitly rely on in our everyday epistemological evaluation as the basis of the concept of justification.

What makes virtue epistemology unique is that it locates the basis of justification within the epistemic agent (Sosa 2009, pp. 187–188). Whether a given belief is justified or not depends on the epistemic ability or virtue inherent to the agent. This epistemic virtue can be paraphrased as a truth-conducive disposition, i.e., a set of properties that leads the agent to the truth in normal circumstances (p. 135). Just as sunflowers are disposed to turn toward the sun and frogs have a tendency to prey on little moving spots, smart and attentive people have a disposition to obtain true beliefs. These dispositions are characteristic abilities inherent to individual sunflowers, frogs, and human beings, respectively. Some of these are innate traits that have evolved through the long history of each species, while others are acquired through postnatal developmental or learning processes; but in either case, they are inherent characteristics belonging to individual organisms. Justification, according to virtue epistemology, is based on a proper manifestation of such individual epistemic capabilities that have taken shape through the historical processes of ontogeny and phylogeny.[9]

Can we extend this line of thought and say that properly trained deep learning models have epistemic virtue? I think we can. By no means does this imply that machines have a personality, or that there is no gap between artificial and human intelligence. "Virtue" here is a technical term denoting specific abilities built into individuals through certain optimization processes, no less and no more. In this technical, restricted sense, it seems perfectly possible to acknowledge epistemic virtue in deep learning models, acquired through optimization processes akin to adaptive evolution and borne out by a set of benchmark tests. For instance, SciNet in the aforementioned work of Iten et al. (2020) can discover physical laws from data, and Alpha Go has the ability to beat a top-level professional *Go* player. The epistemic virtue of a machine is nothing other than the epistemic capabilities possessed by these models. The judgments of models that have such epistemic machine-virtue, then, should be regarded as justified by the standard of virtue epistemology. In other words, when we accept the conclusions of a deep learning model as justified by its truth-conducive capability, demonstrated in benchmark tests, we understand their justificatory status in line with virtue epistemology, that is, in the same sense that we listen to the ruling of a fair-minded judge, the diagnosis of a devoted doctor, or the expertise of an academic authority.

The idea of virtue epistemology that seeks the basis of justification in personal or individual characteristics might appear unscientific and anachronistic to some. Indeed, virtue ethics and virtue epistemology have their roots in Aristotelian natural philosophy, where well-being and excellence are explained in terms of the manifestation of the natural properties inherent to each individual. Hasn't such a metaphysical worldview been overcome by modern science, which has replaced the anthropomorphic conception of nature with mathematical derivations and calculation from universal laws? Wasn't it also the replacement of experts' individual skills and knowledge—that is, their intellectual virtue—with publicly accessible and unequivocally assessable numbers that promoted the development of more objective and democratic decision-making processes in the late modern period (Porter 1996)? If so, our discussion suggests that the theory of machine learning, the cutting edge of contemporary science, is in one sense resuscitating a premodern worldview that we thought had been overcome. This ironic twist, I surmise, is what underlies the sense of amazement, bewilderment, and anxiety people often feel in the face of the impressive successes made by the recent development of machine learning. Deep learning models are surpassing conventional technologies on many fronts, but the way they work remains impenetrable to objective understanding based on universal laws or first principles, which have long been the ideal of modern science. This situation calls for, alongside the scientific as well as engineering endeavor to uncover the fundamental principles governing deep learning models, philosophical reflection on the epistemological nature of such investigations, and on their potential implications for the way science is conducted and conceived. The rest of this chapter explores these philosophical matters, taking as its lead the kinship between deep learning and virtue epistemology put forward in this section.

4.4.3 Philosophical Implications of Deep Learning

Why does deep learning work so well? This is a million-dollar question that is currently being explored by researchers all over the world, and it would certainly be premature to make any judgment at this point. We may, instead, step back and ask from a somewhat detached perspective: what does understanding deep learning models amount to? What kind of discovery or explanation would make us feel we understand them, and what are the implications of such understanding for our concept of knowledge?

From the virtue-epistemological perspective introduced in the previous section, the understanding of deep learning models consists in the elucidation of their virtue or capabilities. Consider how an inquiry into an agent's epistemic virtue would proceed in the case of, say, the study of animal cognition. To elucidate the epistemic ability of some animal, we would first try to identify the relevant functions and then seek their physiological bases. If, for instance, we are intrigued by how and why bats can prey so well in the dark, we should first identify the peculiar epistemic ability—echolocation—that enables them to hunt effectively, and then proceed to examine the physiological structure that realizes this function. In the same way, if we are to understand the performance of a particular deep learning model, we should first identify its epistemic virtue or feature that contributes to its truth-conduciveness, and then ask which network structure is responsible for this ability. This research strategy corresponds to the first item (i.e., the study of model structure) in our list of the three kinds of approaches in machine learning research that we saw earlier (Section 4.3.2).

Specific examples of such structures underlying various epistemic virtues of deep learning models would include convolution, resolution networks (ResNet), recurrent neural networks (RNN), long short-term memory (LSTM), attention, transformer—the list goes on. Identifying and engineering these kinds of mechanisms that contribute to machine-epistemic virtues is the main thread of deep learning research that has led, and is still leading, to progress ever since its birth.

Studies of epistemic virtue aim not only to identify the structure responsible for high performance, but also to understand the reason why that particular structure leads to good performance. For instance, a recent study suggests that deep learning models do their best when the target probability function to be learned is piecewise smooth—that is, when it is composed of smooth differentiable patches, connected by steep cliffs or crevasses where the function is non-differentiable or even discontinuous (Imaizumi and Fukumizu 2019). This kind of theoretical research will help us determine not only the conditions favorable to deep learning models, but also adverse conditions, under which they will not work reliably or effectively. The latter kind of understanding is also important, because deep learning models are known to produce some unexpected fiascoes. The most striking of such a failure is a phenomenon known as an *adversarial example* (Szegedy et al. 2014). It is known that by adding specialized noise unrecognizable to the human eye, one can cause a model to misidentify the contents of an image, e.g., to misclassify a fabricated image of a panda as a gibbon, though the image seems to us no different from the original image of the same panda. This technique may be abused: for instance, one might be able to mislead self-driving algorithms to make wrong decisions by putting specialized stickers on road signs. In order to prevent this from happening, it is important to clarify the mechanism of how and what deep learning models learn from data (Goodfellow, Bengio, and Courville 2016, sec. 7.13).

The goal of this kind of research is to justify the judgments of deep learning models from an external, third-person perspective, by dissecting their epistemic abilities as does an anatomist, as it were. Alternatively, one may consider an internal or first-person perspective strategy that seeks the basis of justification within deep learning models. Ernst Sosa's dichotomy of knowledge will help us see this contrast. According to Sosa, there are two types or varieties of knowledge: one is what he calls *animal knowledge*, and the other is *reflective knowledge*. Animal knowledge refers to the insights or actions made possible by the proper working of the epistemic virtue of a subject. This includes a frog's identifying moving flies as prey, a dog's finding a bone buried in the backyard, and my distinguishing IPA from other ales by taste. These count as justified pieces of knowledge insofar as they are proper manifestations of the epistemic virtue of the frog, the dog, and myself, so that they (we) can legitimately be said to *know* the presence of the prey, the location of the bone, and the kind of beer I am enjoying, respectively. This, however, does not necessarily mean that these cognizers also know how they are able to know what they do. Reflective knowledge requires this kind of second-order understanding, not just of the target fact, but also of "its place in a wider whole that includes one's belief and knowledge of it and how these come about" (Sosa 1985, p. 242). Such reflection may not be necessary for gaining knowledge itself but proves essential if one also wants to know the conditions that guarantee the truth-conduciveness of the belief-generating process that produced the knowledge in question, so that one can guard against potential doubts. Devoid of such reflective knowledge, if someone had asked me why I knew that the drink I'm enjoying is IPA, I would only be able to respond, "because it tasted like that." But a beer-savvy person may be able to explain why it is IPA and not other ales by referring to the way each kind of drink infuses our gustatory and olfactory senses with its taste and aroma under various conditions. Arguably, a person with such knowledge should be considered to have a deeper understanding of beer than I do.

Given this distinction, what kind of knowledge can we attribute to deep learning models? It would be hard to deny that they already have some form of animal knowledge, and the kind of research along the lines mentioned here will no doubt continue to improve this knowledge by identifying its underlying mechanisms. In contrast, it is not evident whether they also have or will acquire reflective knowledge. Do neural nets know not only what they are recognizing and classifying, but also why and how they are making judgments as they do? And are they able to share their rationales or justificatory reasons with us, so that we humans can reflectively understand their decisions? These are exactly the questions being investigated under the recent research trend called *Explainable Artificial Intelligence*, or XAI for short, which studies the accountability or interpretability of deep learning models (Adadi and Berrada 2018). Major goals of XAI include identifying the criteria used by deep learning models in making decisions as well as the rationale that underlies each individual judgment and prediction, and making these open to human users in intelligible ways. For instance, Ribeiro, Singh, and Guestrin (2016) proposed a method for highlighting the regions of images which a model has used to classify objects, while Hendricks et al. (2016) developed a model that explains the basis of its decisions in natural language. These studies are intriguing in that they have the potential to reveal the justification process inherent in deep learning models, and thereby shed light on the aforementioned problem of adversarial examples. The challenge posed by adversarial examples is the possibility that the logic and reasons (called "features") that deep learning models use in making inferences and judgments may be utterly different from those we use in our own reasoning. Such a difference in the way of thinking would make it extremely difficult, if not completely impossible, for us humans to identify and anticipate the circumstances that would cause a model to malfunction. Understanding a model's rationale for its decisions will give us clues as to when such malfunctioning will occur and how to prevent it. A reflective understanding of deep learning models, therefore, is not only of theoretical interest, but is also of great importance in applications.

This discussion suggests that there are two different senses in which we might explicate and understand deep learning models. One is to understand a model's animal knowledge, through an analysis of the characteristic architecture that underpins the model's sophisticated cognitive abilities. Just like biologists and neuroscientists who seek to uncover the physiological basis of cognitive processes in animals and humans, machine learning researchers might seek to identify features of a model that contribute to its truth-conduciveness. We indicated in Chapter 1 that biological species like frogs or human beings are paradigmatic examples of natural kinds, characterized by the physiological properties and behavioral tendencies peculiar to each species. It is by classifying a particular "thing" in front of us into a natural kind such as a frog or human that we can predict and understand its behavior, such as that it will spawn eggs in a pond in the spring or that it will display a certain reaction if such-and-such a drug is administered. Physiological studies in general are aimed at elucidating the making and working of such biological natural kinds. Likewise, a deep learning model instantiates a probabilistic kind, a natural kind that specifies a particular pattern of inductive problems (see Section 4.3.1). Just as biological natural kinds like frogs and humans solve cognitive problems that arise in their respective environments, deep learning models are probabilistic kinds that cope with cognitive tasks of their own. And just as frogs and humans have different cognitive features and environments, different deep learning models like GoogLeNet and Alpha Go have distinct cognitive architectures and flourish in different "habitats." At the same time, different models are not entirely different: there are some versatile modules adopted in a wide range of models built for different purposes. These universal features are like a machine-analog of *homologies*, common biological traits or structures observed in distinct taxa due to their shared evolutionary origin. Examining how these commonalities as well as specific features contribute to the performance of deep models, then, will define the first category of deep learning research. This line of research follows the same path as other natural sciences, progressing through the identification and examination of relevant natural kinds.

On the other hand, understanding the explainability and interpretability of deep learning models will require a different research strategy. The focus of this second research category is the reflective knowledge of deep learning models, i.e., knowledge about how they reason and on what ground they make judgments. Whether such knowledge is attributable to machines should be determined not by simply observing their behavior from without, but rather from within, by "putting ourselves in their shoes." Taking a zoological analogy again, this is akin to pondering how an animal, say a bat, cognizes its environment, and trying to understand "what it is like to be a bat." As is suggested by the recent studies on *representation learning*, deep learning models are seeing the world in their own way—that is, they are extracting patterns from given information and making full use of the acquired patterns for solving cognitive tasks. A striking example is an image recognition model that developed a neuron that specifically responds to the image of cats, just by learning from data. This and other cases suggest that deep learning models are able to discover what we would call natural kinds on their own, and use them to structure the world. As it happens, building good representations is known to improve a model's generalization capability and to facilitate the application of a model trained in one domain, say image classification, to problems in another domain, say text processing (a procedure known as *transfer learning*). Given that the role of natural kinds is to secure a foothold for inductive reasoning and extrapolation across different sense modalities by carving nature at its real and objective joints, this finding buttresses our claim here that the patterns identified by models are indeed natural kinds. If so, and if we consider that natural kinds furnish the basic building blocks of our worldview, understanding the representations/natural kinds used by machines proves to be an essential step toward understanding how deep learning models actually *think*. Do they carve nature as we do, or are they seeing completely different natural kinds? In Section 4.2.4, we observed that the real patterns that cognizers discern may depend on their perceptual sensitivity or amount of data available. In view of this, there is little reason to expect that the patterns in the world discovered by deep learning models trained with an enormous amount of data coincide exactly with those that we tend to consider real. The question, then, is the extent and upshot of the discrepancy. If the natural kinds used by deep learning models are utterly different from ours, can we ever understand their epistemic process reflectively? And how can we even know whether they use the same natural kinds or not in the first place?

Philosophically minded readers might notice a similarity between the problem we encounter here and Quine's infamous *indeterminacy of translation* (Quine 1960). This thesis puts into question the commonly held idea that there is a uniquely correct translation rule between two different languages: instead, there are always multiple (possibly infinite) translations that are equally good but mutually inconsistent. Quine's argument unfolds as follows. Imagine that you are a field linguist doing research in a tribal community which previously has had no interaction with the rest of the world, and whose members speak a language utterly unknown to us. Living with those people, you find that they utter "gavagai" upon seeing a rabbit. Then you reason that "gavagai" in their language should mean something like "Lo, a rabbit." in English. That, however, is not the only possibility. Maybe "gavagai" should be translated into a more metaphysically perverse sentence like "Rabbithood is instantiated over there." Or maybe what we mean by rabbits is believed to be reincarnations of ancestors in that tribe, so the native people may be saying that their ancestors are visiting them. To resolve this ambiguity and determine the meaning of "gavagai," we need to take into account the rest of their linguistic activities as a whole.

However, Quine argues, there is no guarantee that your best efforts to survey their entire linguistic activities will converge to a unique translation manual. Other field linguists may come up with a completely different translation manual, according to which "gavagai" means an ancestral visit. In this way, "manuals for translating one language into another can be set up in divergent ways, all compatible with the totality of speech dispositions, yet incompatible with one another," or so argues Quine (1960, p. 27).

A similar problem of *radical translation* may arise between machines and us. As an example, recall the node in the aforementioned image classification model that fires exactly when the image contains a cat. Does this model really recognize what we call a cat? Maybe so, or maybe it is just responding to the combination of cat-like whiskers and ears, or thinking that "cathood is instantiated over there," and so on—the list could be continued indefinitely.[10] Note that this is *not* a problem of underdetermination, where we are unable to decide which among multiple possible translation rules is the correct one due to insufficient data. Quine's point is more radical and suggests the possibility that there may not be an objectively "correct" translation rule, or even a "true semantic content" which translations aim to reveal. If this is the case, then XAI's attempt to translate the "thoughts" of deep learning models into our natural language would be a quest with no definitive answer. Of course, even without an objective answer, we may still tell a story about how they reason. That, however, will be just one interpretation, and there can be multiple interpretations that are equally good but mutually incompatible.

This casts serious doubt on the search for the explainability of deep learning models and their reflective understanding. If we can never know "what AI is really thinking," why bother prying into it? We can confirm through benchmark tests that deep learning models have high performance/animal knowledge—what else do we want? Skilled researchers and engineers may well be able to get an idea about a model's performance by glancing at its benchmark scores or other indices, and if necessary, the source code. It is understandable, then, that they do not feel any need for explainability, particularly if by itself it does not help them improve the performance of their models.

The story changes, however, when it comes to applications. When applying a particular model to solve any concrete problem in, say, business, policy making, social planning, or scientific research, one always needs to convince the stakeholders (one's boss, government officials, politicians, reviewers, etc.) of the appropriateness of the method. And as Porter (1996) has pointed out, a demand for accountability always arises from those external parties. From a practitioner's standpoint, what matters most is not some esoteric details, but whether a given deep learning model is effective in the task at hand, and possibly more importantly, whether it is free of unintended glitches. This is nothing but the problem of justification. Machine learning researchers may respond to such worries by citing a model's benchmark scores or past successful cases, trying to assure their clients that these scores confirm the model's truth-conduciveness. But to the eyes of practitioners, these scores are nothing more than currencies within the machine learning community. What matters to them is whether the method is safe and effective in this or that particular task. This is Humean skepticism, and for that very reason one cannot preach it down logically; moreover, it cannot be dismissed as an unfounded fear either, given the actual reports of adversarial examples.[11] The only thing one can do is tell a story which would describe how and what kind of assumptions may possibly mitigate the skepticism. Traditional statistics aimed to answer the skepticism by framing a given inductive problem in terms of a natural kind (statistical model) and by deriving various estimation methods from it (Chapter 1). Since this posited natural kind remains a conjecture and can never be fully confirmed, and the assumption itself is deemed to be "wrong" anyway (Box, Luceño, and Paniagua-Quinones 2009), this in fact is a sort of metaphysical story. But at least it has proven to be a useful story, in that it explicates the conditions that would justify our inductive practices. On the other hand, the prohibitive complexity of both the natural kinds and tasks in deep learning precludes any hope for such deductive explanations. Under this situation, the explainability of deep learning models will hopefully provide a means to respond to the external demand for accountability. For this reason it cannot be easily dismissed, even if it turns out to be a difficult endeavor that lacks an objective, uniquely correct answer.

Further Reading

Anderson (2008) is an advanced undergraduate or graduate-level introduction to the theory behind AIC and other information criteria. See Konishi and Kitagawa (2008) for a more theoretical exposition. The philosophical implications of AIC have been explored by Forster and Sober (1994) and Sober (2008). Kelleher (2019) and Krohn, Beyleveld, and Bassens (2019) are accessible guides to the core ideas behind deep learning that require little mathematical background, while Goodfellow, Bengio, and Courville (2016) is a standard reference with theoretical details. Pragmatist epistemology is championed by Stich (1990). Virtue epistemology is succinctly summarized by Battaly (2008).

Notes

1. One might find it odd to talk about a "prediction" of what is known to have happened. However, one can pretend as if the actual data has not been observed, and consider which parameter values would best predict the outcome that happened to obtain in the real world. The likelihood measures the predictive ability of a hypothesis in this counterfactual sense.
2. For a model $M(\boldsymbol{\theta})$ characterized by the parameters $\boldsymbol{\theta}$, the MLE given some data $\boldsymbol{x}$ is $\hat{\boldsymbol{\theta}} = \arg\max_{\boldsymbol{\theta}} P(\boldsymbol{x}; M(\boldsymbol{\theta}))$, and the model's maximum likelihood is $\ell(M) = P(\boldsymbol{x}; M(\hat{\boldsymbol{\theta}}))$.

3. Premature, because we did only 10 tosses. The MLE converges to the true value as the sample size increases. It also has other desirable properties, like asymptotic normality, optimality, and efficiency (Wasserman 2004, sec. 9.4).
4. The method is often called "regression" when the predicted variable Y is continuous, and "classification" when it is discrete. But in this book, we use "regression" for both cases.
5. Depending on the context, explanatory variables are also called *independent variables, regressors, covariates*, etc., while response variables are called *dependent/objective variables, regressands, targets*, etc. Although independent/dependent in particular is a common moniker, we avoid this terminology because it may erroneously suggest that explanatory variables must be probabilistically independent from each other.
6. Of course, this does not mean that MLEs cannot be tested.
7. A particularly important assumption is that the set of candidate models contains the true model (i.e., the probability model itself). This condition is dropped in a more general criterion (called TIC) studied by Takeuchi (1976) and Stone (1977).
8. Adaptive evolution is often likened to hill-climbing on a fitness landscape, where the fitness of an organism is determined as a function of its traits and environment. In deep learning, fitness corresponds to a model's likelihood or predictive accuracy, while traits correspond to parameters. Natural selection, then, is a kind of optimization process of adjusting traits (parameters) to increase fitness (likelihood). I analyzed this analogy between biological evolution and machine learning elsewhere (Otsuka 2019).
9. This naturally leads to the question of what counts as a "proper" manifestation of a capacity. Sosa analyzes this notion in terms of his concept of aptness. Another approach would resort to the evolutionary history of the cognitive trait and relate it to what Millikan (1984) calls proper function. But we will not go further into this problem here.
10. Indeed, a recent study suggests the possibility that many deep learning image recognition models are tacitly using background information in object classification (Xiao et al. 2020). Hence, it is a real possibility that what they call a "cat" turns out to be what we call "a typical background of cat pictures."
11. Indeed, adversarial examples can be thought as a real-world case of Wittgenstein's notorious rule-following paradox.

5
CAUSAL INFERENCE

As we have discussed so far, modern statistics has approached the problem of induction from a probabilistic perspective, framing it as a matter of predicting unobserved phenomena via what we have called probabilistic kinds. But prediction is not the sole aspect of inductive reasoning. In this chapter we change gears and take an alternative, causal, perspective. Causal relationships are part and parcel of everyday as well as scientific inductive reasoning. If we predict, for instance, that vegetable prices soar after droughts, that is because we believe that precipitation affects crop growth. In this way, we often resort to a causal relationship to deal with an inductive problem. Indeed, Hume notoriously blurred the distinction between induction and causality, taking the justification of inductive reasoning and the confirmation of causal relationships as two sides of the same problem. This may sound like comparing apples and oranges to students who have taken a modern statistics course, in which we are routinely warned that correlation is *not* causation. The admonition is of course well-taken and true; but nevertheless, probability and causality are not *irrelevant* to each other, and identifying their exact relationship is the main focus of *statistical causal inference*, which has made significant progress over the past half-century in developing various methods for inferring unknown causal relationships from observational data alone, without resorting to experiments. What is the true nature of this triadic relationship between induction, probability, and causality? This chapter addresses this question by taking the same approach we have adopted throughout this book, that is, through semantic, ontological, and epistemological analysis. What does it mean for something to cause something else? What kind of ontological assumptions are implied in such a statement, and what kind of method would enable us to know such entities? Through these questions, this chapter aims to expose the philosophical implications of statistical causal inference.

5.1 The Regularity Theory and Regression Analysis

The canonical place to begin an inquiry into causality is, again, Hume. As mentioned previously, for Hume the problem of induction *was* the problem of causality. What, then, is causality? According to Hume, it is a relationship that satisfies the following three conditions:

1. The cause and effect are *spatiotemporally contiguous* to each other.
2. The cause *temporally precedes* the effect.
3. There is a *constant conjunction* between the cause and effect.

When a cue ball hits another ball at rest and sets it in motion, the two balls must be spatiotemporally contiguous at the moment of collision, and the motion of the cue must precede that of the other. Moreover, this phenomenon—in which a ball at rest is set into motion after collision—is observed constantly and repeatedly. Hume claimed that these three conditions are all there is to causal relationships. In particular, he denied any further specification of causal relationships as redundant and meaningless, such as the commonly held opinion that the cause must have some sort of "power" of bringing about its effect, or that there is a "necessary relationship" between the cause and effect. These common ideas frequently appear in our causal talk, but according to Hume, they have no empirical basis. All we observe in causal relationships is a constant succession of effect-event after cause-event; we never observe anything like the "power" of a cause that allegedly makes the effect happen, or any kind of "necessity" between the two events. Hence, insofar as we remain within the realm of data and experience, as Hume thought we should, causal relationships are to be characterized by the above three conditions, and nothing more.

The Humean conception of causality sketched above falls under the rubric of the *regularity theory* of causation. A causal relationship, according to this theory, is nothing but a certain sort of regularity between events. One way to cash out this regularity is to resort to the notion of statistical dependence between random variables (Pearson 1892). As we saw in Chapter 1, random variables X and Y are dependent if there are values x, y of these variables for which $P(x|y) \neq P(x)$. Let us denote the independence of random variables X and Y under probability function P by $X \perp_p Y$, and their dependence by $X \not\perp_p Y$. Then Hume's condition of constant conjunction can be written as $X \not\perp_p Y$. But even if we put aside the other two conditions, this alone does not suffice to establish a causal relationship between X and Y, for the dependence may also arise due to a common cause of the two. Lynda always suffers from headaches when the scale on her barometer is low, that is, Lynda's headaches and the barometer reading are correlated; but this is just a spurious correlation caused by a common cause, atmospheric depression. Such spurious correlations can be ruled out by conditioning on the common cause, also called a *confounding factor* or *confounder*. If it is atmospheric depression and not her glancing at the barometer that is causing Lynda's headache, then no correlation will be observed between the barometer reading and Lynda's headache on days with the same pressure. Denoting the confounding factor (here, the atmospheric pressure) by Z, this means

$$P(x|y, z) = P(x|z)$$

holds for all values x, y, and z. If this holds, we say that X and Y are *conditionally independent* given Z, or that Z screens-off Y from X, and write $X \perp_p Y | Z$. With this, we can tentatively rephrase the regularity theory as the claim that there is a direct causal relationship between X and Y if and only if $X \not\perp_p Y$ and there is no confounding factor Z such that $X \perp_p Y | Z$. This definition of causality is reductionistic, in the sense that it aims to replace causal relationships with probabilistic language. That is, it defines causal relationships to be nothing more or less than a sort of probabilistic relationship of dependence. Then, in order to understand causality we won't need to introduce any further "entities" beyond our familiar probability models defined in Chapter 1. In this way, the regularity theory reduces the concept of causality to the concept of probability, both semantically and ontologically.

The ontological reduction, if possible, is good news for the epistemological project of discovering causal relationships. For if causal relationships are nothing but a sort of probabilistic relationship, no more than familiar statistical methods are needed to identify them. The foremost candidate for such methods is regression analysis. In common practice, a regression model of a given target variable Y takes the putative causes and confounding factors $X_1, X_2, \ldots, X_n$ of Y, often called *covariates*, as the explanatory variables. The simplest linear regression model that ignores interactions among causes will have the following form:

$$y = \beta_1 x_1 + \beta_2 x_2 + \cdots + \beta_n x_n + \epsilon,$$

where ϵ is an error term that follows some probability distribution, and each parameter β_i gauges the impact of the ith factor X_i on the effect Y. The parameters can be easily estimated using standard regression analysis. In particular, if one of the parameters is inferred to be zero (that is, the hypothesis that it is zero has a high posterior probability, or cannot be rejected by a statistical test), we may conclude that the corresponding factor is not a direct cause of Y. The advantage of regression analysis is that it can simultaneously address the problem of confounding. Incorporating a variable, say X_i, into a regression model amounts to conditioning on X_i. Hence, even if X_i confounds Y and X_j (that is, even if X_i is a confounder between Y and X_j), an estimate of β_j will reflect the (partial) correlation relationship between X_j and Y free from the effect of X_i, that is, the statistical association that remains after X_i is held fixed. We noted above that a causal relationship between X and Y according to the regularity theory requires $X \not\perp_p Y$ and the absence of a confounding factor Z such that $X \perp_p Y|Z$. Hence, if one wants to know whether X_j causes Y in this sense, all one has to do is run a regression analysis that includes all potential confounding factors in the covariates, and see if β_j is zero or not.

Regression analysis has been one of the standard methods for the study of causal relationships, especially in cases where direct experimentation is infeasible or difficult. The carcinogenic risk of smoking, for instance, has been confirmed by regression models that consider, in addition to smoking history as the primary explanatory variable, various putative confounding factors such as age, family, occupation, genetic factors, and so on. Records of such actual applications may be taken to vindicate the regularity theory as *the* correct understanding of causality. However, the theory has also been challenged on several fronts. The first issue is that regression analysis addresses just one among the three Humean conditions, namely constant conjunction, and says nothing about spatiotemporal contiguity or temporal succession. Since the probabilistic relation of dependence is symmetric as we saw in Chapter 1, a nonzero regression coefficient alone does not tell us whether X causes Y or vice versa. Even if we set aside this problem of directionality, the identification and selection of covariates poses practical as well as conceptual challenges. In most cases, it is not evident at all which factors should be included in a model as covariates. In order to correctly identify a causal relationship, the covariates must cover all the possible confounders; but what can confound, say, smoking history and the risk of lung cancer? Even with our most painstaking and extensive search for confounders, there may still be ones that are out of reach of even our wildest imagination; and for that reason, any conclusion from a regression model must remain at best tentative. Moreover, there are variables that should not be included in a regression model. If the variables of our interest Y, X respectively have distinct causes C_Y, C_X, which in turn have a common effect E, so that the causal structure has the form $Y \leftarrow C_Y \rightarrow E \leftarrow C_X \rightarrow X$ in the language of causal graphs we will see shortly, then the middle E should *not* be included as a covariate in a regression model. This kind of causal structure is called an M-structure (Pearl 2009); in such a case, conditioning on E creates a spurious correlation between Y and X (see Section 5.3.1 for an explanation of why this happens).

The difficulty of covariate selection is well-illustrated by the infamous phenomenon of *Simpson's paradox* (Simpson 1951), which vividly shows that the statistical association between two variables may change or even be reversed depending on the variables one conditions on. Let us explain this with an actual study of gender bias among graduate school admissions to UC Berkeley. In 1973, about 44% of male applicants to the university were admitted, whereas the admission rate for female applicants was just 35%, which is significantly lower. This led to the suspicion that the university was discriminating against women. A closer look at each department, however, revealed that in the majority of departments, women were more or at least equally successful compared to men. At first sight this is puzzling: female applicants fared better (or at least not worse) than male applicants in each department, but worse at the entire university level. This apparent paradox, however, is no real paradox; the truth was that women tended to apply for relatively competitive departments with few seats, which resulted in a larger proportion of women who were not admitted in the entire pool of applicants (Bickel, Hammel, and O'Connell 1975). Turning our eyes to the causes of this apparently paradoxical phenomenon, we observe that in this case the gender X affected admission Y in two ways. One is a direct influence, where being a woman contributed positively to admission (because women tended to fare better than men in admission). The other is an indirect influence through the applied department Z, where being a woman negatively affected admission (because women tended to apply for competitive departments). Hence, if we want to judge whether there was discriminatory practice, i.e., know whether gender had a direct influence on admission, we should condition on the applied department Z to remove the second (indirect) effect. This diagnosis, however, presupposes that we already know the aforementioned causal mechanism, or at least have it in mind as a hypothesis. In the absence of such foresight, or if there are yet other unknown factors, there is no telling which variables should be included as covariates.

All of what we have seen here is an epistemic problem. What is more, covariate selection also raises a conceptual challenge to the regularity theory. We noted that the reductionistic definition of the regularity theory identifies a causal relationship with a certain kind of probabilistic relationship. For such a reductionistic enterprise to succeed, the definiens must not involve any causal concepts, for otherwise the definition cannot claim to have reduced causality to probability, or replaced the former with the language of probability. What we have just seen, however, is that an appropriate selection of the covariate Z, which plays a crucial role in the regularity theorist's definition of causality, presupposes knowledge of the very causal relationship being defined. For one thing, the fact that Z is or is not a confounder, or part of an M-structure, is nothing but a causal feature of the variable. The regularity theory's characterization of causal relationships, then, hinges on the very notion it tries to define, and for this reason it cannot claim to have successfully defined causality solely in terms of probability. The failure of this semantic reduction suggests the possibility that causal relationships are not probabilistic relationships, or what we have been calling probabilistic kinds, after all. This is not just a possibility—as a matter of fact, causality is *not* probability. But what is it then? To see this, we need to consider counterfactual situations—in other words, not just the actual world, but also possible worlds.

5.2 The Counterfactual Approach

5.2.1 The Semantics of the Counterfactual Theory

To explore an alternative strategy to the regularity theory, let us think about what we really mean when we make causal claims. Causal statements are ubiquitous both in our everyday life and scientific contexts. Statements like "an asteroid impact caused the extinction of dinosaurs," or "you got a cavity because you ate too many candies" are typical examples of statements that claim causal connections among the events involved. In so claiming, we are not necessarily trying to establish a regular pattern between the relata. For one thing, it won't make much sense to talk about a regular connection between the meteorite impact and the extinction of the dinosaurs, which presumably happened only once in the earth's history. Rather, what we mean when we make these causal statements should be something like: "Were it not for the asteroid impact, the dinosaurs would have continued to thrive," or "If you had refrained from eating candies, you would not have had a cavity." These are counterfactual statements, some of which we have already encountered in Chapter 3. Counterfactuals, as we saw there, have us imagine how things would have developed in non-actual situations, and in this respect they are to be distinguished from ordinary conditionals of the form "if A then B," which deal with regularities holding in the actual world.

The philosopher David Lewis, whom we met in Chapter 2, claimed that such counterfactual thinking is what characterizes causal relationships, and he proposed the *counterfactual theory of causation* (Lewis 1973).[1] According to this idea, *E causally depends* on *C* when the following two counterfactual conditions hold:

(L1) If *C* were true, *E* would also be true.
(L2) If *C* were not true, *E* would also be non-true.

Also, if there is a finite sequence of events $D_1, D_2, \ldots, D_n$ between *C* and *E* such that each term causally depends on its predecessor, i.e., D_1 causally depends on *C*, D_2 on $D_1, \ldots,$ and *E* on D_n, then *C* is said to be a cause of *E*. In our following analysis of causation, however, we will ignore this kind of sequence and focus on the causal dependence between just two events, stipulated by conditions (L1) and (L2).

Immediately noticeable in Lewis's definition of causality is its affinity to Nozick's two counterfactual conditions of knowledge that we saw in Chapter 3. Indeed, just like in Nozick's treatment, the truth conditions for Lewis's counterfactual conditions (L1) and (L2) are stipulated in terms of possible world semantics (Section 3.3.2). In particular, (L1) "If *C* were true, *E* would also be true" is true in the actual world when either

(i) *C* does not hold in any possible world, or
(ii) both *C* and *E* hold in some possible world which is closer to the actual world than any possible worlds in which *C* holds but not *E*.

The condition (i) is a technical proviso and can be ignored here. More substantial is (ii). When (L1) is making a counterfactual speculation, by assumption *C* does not hold in the actual world. But (ii) demands that there is some possible world in which *C*, and also *E*, hold. Call such a world an *exemplary-world*. Not all possible worlds are exemplary; there are worlds where *C* holds but not *E*. Call them *counter-worlds*. But as we noted in Section 3.3.2, we won't have to count these worlds as counterexamples if they are too distant from our actual world. The truth condition (ii) of the counterfactual statement captures this intuition and declares (L1) to be true if the exemplary-world is similar to the actual world than any counter-world. Likewise, the truth condition of (L2) can be stipulated by reversing this argument, that is, by replacing *C*, *E* above with their negations ¬*C*, ¬*E*.

Let us apply Lewis's definition to a causal statement, "Eating sweets causes a cavity." For this to be true, (L1) must hold—i.e., for a person who does not have a sweet tooth, the counterfactual "If she had eaten a lot of sweets, she would have had a cavity" must obtain. This requires that there is an exemplary-world in which she eats a lot of sweets and gets a cavity. Yet there may be other possible worlds where she is fortunately free from cavities despite her high-sugar diet. But suppose that these worlds are rather different from the actual one, in that, say, she brushes her teeth really carefully, or fluoride has been added to tap water. In this case, these worlds should not be counted as counterexamples, and (L1) is judged to hold. The next requirement is (L2), that is, for a person who does have a sweet tooth, it must be the case that "Had he not eaten lots of sweets, he would not have had a cavity." The counter-world in this case is one where that person unfortunately gets a cavity despite his sugar-free diet. But suppose again that these worlds are all quite different from ours, in that he does not brush his teeth at all, or cavity-causing bacteria are much more virulent, while the exemplar-worlds in which he does not get a cavity are very much like the actual world except for his preference for sweets (and the absence of cavities). If that is the case, (L2) holds. And if the two counterfactual conditions (L1) and (L2) both hold in this way, eating sweets is deemed to be a cause of cavity.

Although the counterfactual theory may look a bit complicated at first sight, it seems to capture well our intuitions about causality; that is, it seems to give a good semantics of causal statements (readers are invited to try it out with some examples of their own).[2] At the same time, however, it gives rise to an epistemological problem. For one thing, we are tied to the actual world by definition, and are never able to peep into other possible worlds and check how things are going there. If so, how can we *actually* confirm the truth or falsity of the conditions (L1) and (L2), and judge one event to be a cause of another? Problematically, the counterfactual theory remains silent to this question—it may provide us with a good semantics of causation, but no epistemology. In what follows, we will take up this epistemological challenge and search for statistical methods that would allow us to probe into possible worlds.

5.2.2 The Epistemology of Counterfactual Causation

Statistical Tests and Causal Inferences

Is there a way to explore possible worlds using data available in the actual world? One possible clue to this question might be found in our discussion of hypothesis testing in Chapter 3. There, we observed that statistical hypotheses represent possible worlds, and that a test provides us with an empirical procedure for deciding which possible world is the actual world we live in. And as we previously noted, Nozick's two counterfactual conditions that embody the logic of statistical tests have a close resemblance to the Lewisian definition of counterfactual causality. This suggests that we pursue the possibility of making use of a method similar to statistical testing to determine the truth value of causal statements.

To spoil the conclusion, this idea does not work out. But it is not wide of the mark either. Statistical tests and causal inferences do have something in common; indeed, one could even say that *statistical tests are a kind of causal inference*. Since identifying the commonality and difference between the two will serve to highlight the distinctive nature of causal inference and the challenges it poses, it will be worthwhile to make a brief detour here and look at their subtle relationship.

Let us begin by recalling that a statistical test is nothing but a function which, given data, determines whether or not we should reject a null hypothesis (Section 3.2). Let R_0 stand for the rejection of the null hypothesis, and H_1 for the proposition that the alternative hypothesis is true (i.e., the null hypothesis is false). What we can confirm from the data is either R_0 or $\neg R_0$, and from this information we want to judge whether or not H_1 holds. In order to make this judgment in the classical statistics framework, we need to evaluate the reliability of the testing process. The reliability is measured by the extent to which Nozick's two conditions are satisfied:

(N1) If H_1 were true, it would be the case that R_0 (i.e., the null would have been rejected).
(N2) If H_1 were not true, it would not be the case that R_0 (i.e., the null would not have been rejected).

The extent to which these conditions are met is numerically gauged by the power and confidence coefficient of a test, respectively. Comparing these two conditions with Lewis's definition of counterfactual causality, we note that they are nothing but the conditions for H_1 to be a cause of R_0. Hence, the following equivalences hold: a test is reliable ⇔ the Nozick conditions hold ⇔ the Lewis conditions hold ⇔ H_1 is a cause of R_0. This leads us to conclude that the condition of a reliable test is that the test establishes a strong causal relationship between H_1 and R_0, so that if the cause H_1 holds, so would the effect R_0; and if H_1 does not hold, neither would R_0. This reinterpretation allows us to identify a statistical test with a kind of causal inference that infers the presence or absence of the cause H_1 on the basis of the observed effect R_0 or $\neg R_0$.

The analogy between a statistical test and an inference to causes should become more intuitive if we consider the following example. Consider a fire alarm, whose purpose, of course, is to notify the occurrence of fire by emitting a warning sound. When the alarm goes off, we infer the presence of a fire, and otherwise we assume everything is normal. It is intuitively evident that the reliability of this inference depends on the strength of the causal relationship between fire and the alarm. A good alarm is one that goes off when, and only when, there is a fire. This requirement is met when the device establishes a strong causal connection between fire and the sounding of the alarm. A statistical test is like a fire alarm, which aims to establish a strong causal connection between the fact (H_1) and the rejection (R_0) of the null hypothesis, instead of fire and the alarm sound. In both cases, it is the presence of a stable causal process that allows us to safely trust the process.

A statistical test, therefore, can be regarded as a kind of causal inference that infers a cause from a given effect.[3] The crux of testing theory is to make this inference as reliable as possible, by constructing a firm causal connection between them. But the very point of "construction" reveals the inadequacy of statistical tests for the present purpose of confirming a causal relationship. Why? Because our objective, recall, is to determine from the data whether or not the Lewisian conditions (L1) and (L2) hold. That is, the presence or absence of a causal relationship between C and E is yet unknown—it is the very target we are trying to know. By contrast, testing theory begins by *presupposing* a causal relationship in the form of a statistical model. Using a certain test method is tantamount to assuming a certain causal connection between the facts and judgments. On the basis of this connection, one infers the truth or falsity of the hypothesis as a "cause" from the decision of rejection as an "effect." And, as we have discussed in Chapter 3, testing theory remains silent as to the veracity of this assumption, or at least, it does not assess its veracity vis-à-vis observed data. All that tests do is infer a cause from an effect by virtue of a causal relationship that is posited *a priori*; it does not examine the presence or absence of this causal relationship itself. But our present concern is exactly this latter question, i.e., whether there is a causal relationship between one event and another to begin with. "Presupposing" such a relationship as tests do is just begging the question. What we are after is a methodology that infers this relationship from the data without assuming it; and for that we need to search somewhere other than statistical testing.

Potential Outcomes and Randomization

So let us shift gears and search for another path. Maybe we should first try to identify what kind of data would be required in order to verify Lewis's two conditions (L1) and (L2). To make things specific and easy to imagine, we stick to our example in this section and let X stand for "eating sweets" and Y for "having a cavity." Each subject in our dataset is observed to be either X or $\neg X$, and either Y or $\neg Y$. What we want to know is the answers to the following two counterfactual questions:

1. Would those observed to be $\neg X$ (not eating sweets) have been Y (have a cavity) if they were instead X?
2. Would those observed to be X (eating sweets) have been $\neg Y$ (have no cavity) if they were instead $\neg X$?

If we can give a positive answer to these questions for a large proportion of the subjects, (L1) and (L2) would be considered to be satisfied, and we could conclude that X is indeed a cause of Y. The problem, however, is that answers to these two questions can never be determined empirically, due to their counterfactual nature. If one person was observed to be X, then our epistemic access to a possible world in which that person is $\neg X$ is barred forever. You may well be able to *imagine* a world in which I don't like sweets, but given that I *do* like sweets in the actual world, you, or any one of us who lives in this world, cannot *observe* that possible world. It then appears that causal inferences are trying to do the impossible, by asking questions that can never be answered within the confines of the actual world. This apparently insurmountable difficulty has been called *the fundamental problem of causal inference* (Holland 1986). Paraphrased philosophically, the problem points to the metaphysical impossibility of observing a possible world from the actual world.

But we should try to press forward rather than get stuck in a metaphysical swamp. Let us consider X and Y to be random variables with possible values 1 and 0, standing for true and false, respectively. We further introduce another set of random variables Y_0 and Y_1, where Y_0 represents *the value of Y an individual would exhibit if that individual had been registered $X = 0$* (e.g., the presence or absence of a cavity if one were not a sweets lover), and Y_1 represents *the value of Y an individual would exhibit if that individual had been registered $X = 1$* (the presence or absence of a cavity if one were a sweets lover). Y_0 and Y_1, therefore, denote potential states of Y that would be realized depending on the status of X, and for this reason are called *potential outcomes*. Note that the value of Y_0 can be observed only for those who are registered $X = 0$; for others (individuals

TABLE 5.1 For each subject, only one of two potential outcomes Y_0, Y_1 can be observed; the value of the other is always missing (denoted by "–").

Subject	A	B	C	D	E	...
X	1	0	0	1	1	...
Y_0	–	0	1	–	–	...
Y_1	1	–	–	1	0	...

with $X = 1$), it is only defined and never observed. Philosophically speaking, the value of Y_0 for individuals with $X = 1$ is realized only in some possible worlds different from ours; statistically speaking, it is always *missing* in the actual world. In general, for each person we can observe at most either Y_0 or Y_1, but never both; there is no way to identify the value of Y_0 for $X = 1$ individuals, or that of Y_1 for $X = 0$ individuals (see Table 5.1). With this setup, the fundamental problem of causal inference can be paraphrased as the problem that one of the two potential outcomes always has a missing value.

Using the potential outcome notation, the fulfillment of Lewis's two conditions for a particular person (so that that his/her being x is indeed a cause of his/her being y) can be expressed as $Y_1 = 1$ and $Y_0 = 0$. In other words, $Y_1 - Y_0 = 1$, for that person. The mean of this difference

$$\mathbb{E}(Y_1 - Y_0) = \mathbb{E}(Y_1) - \mathbb{E}(Y_0) \tag{5.1}$$

then represents the extent to which the causal relationship in question holds in a population. This is called the *average treatment effect*, and if it is closer to 1 we can think that X is indeed a cause of Y in the population. However, since at least one of Y_0 and Y_1 is always missing for any individual for the reason discussed above, the aforementioned mean (Equation 5.1) cannot be calculated or estimated directly from data. What we *can* estimate are the mean of Y_1 given $X = 1$ and the mean of Y_0 given $X = 0$, simply by averaging the values of Y for those individuals observed to be $X = 1$ for the former and $X = 0$ for the latter. Hence, the difference of the conditional expectations

$$\mathbb{E}(Y_1|X=1) - \mathbb{E}(Y_0|X=0) \tag{5.2}$$

is estimable from the data. If, therefore, this conditional mean (Equation 5.2) matches the unconditional mean (Equation 5.1), we may circumvent the fundamental problem and infer causal effects from observed data.

Under what circumstances, then, do they coincide? Evidently it is when $\mathbb{E}(Y_0) = \mathbb{E}(Y_0|X=0)$ and $\mathbb{E}(Y_1) = \mathbb{E}(Y_1|X=1)$. This holds true if X and Y_i are independent, for $i = 0, 1$ (Chapter 1).[4] Unfortunately, however, there is no reason to expect this independence to hold. The latter independence $P(Y_1) =$

$P(Y_1 | X = 1)$, for example, claims that the probability that actual sweets lovers develop a cavity is equal to the probability that a random person *would* have a cavity *if they had been* a sweets lover. But those who like sweets do not just regularly eat sugary foods; they may also tend to have additional unhealthy dental habits like drinking soda or frequent snacking between meals. In the presence of these confounding factors, the chance of having a cavity among those who are actually confirmed to eat sweets, $P(Y_1 | X = 1)$, may be higher than the chance that randomly chosen people who are "turned into" sweets lovers develop a cavity in a possible world, namely $P(Y_1)$. If so, X and Y_1 are not independent, and given the ubiquity of potential confounding factors (see the discussion in the previous section), we can't just assume the equality of the above two means (Equations 5.1 and 5.2).

Is there a way out of this impasse? Yes. One way is to conduct an experiment in such a way that *makes* the variables independent. We can imagine, for example, an experiment in which a coin is tossed for each subject, and if it lands on heads, we ask the subject to eat sweets every day, and if it lands on tails, we ask them to abstain from any sugary food. Since the coin toss is random, X in this experiment would be independent from any other variables, including Y_0 and Y_1. This is the core idea of Fisher's famous *randomized control trial*, or RCT for short. RCT randomly decides whether subjects undergo the treatment in question (in this case, eating sweets every day) or go into the control group, and then compares the means of the target variable (the rates of cavity) between the treatment/control groups. If the difference is significant (Chapter 3), the treatment is concluded to have a causal effect. The rationale of this reasoning is that the randomization makes the treatment X independent of the potential outcomes Y_0, Y_1, thereby closing the gap between, on the one hand, the between-group difference (Equation 5.2) estimable from actual observation, and on the other, the average treatment effect (Equation 5.1) defined only over possible worlds.

RCT has been the royal road of causal inference, and much of what we take to be scientific knowledge today is based on it. Most newly developed drugs, for instance, must pass an RCT experiment before they are approved for marketing. But it is by no means a foolproof method (Worrall 2007), and conducting an RCT trial involves various pragmatic as well as ethical challenges. RCT experiments usually call for meticulous preparation and a large amount of human as well as financial resources. In addition, the requirement of random assignment raises serious ethical concerns in some cases. For example, it would be ethically impermissible to force randomly chosen people to smoke cigarettes in order to confirm the health risk of smoking. There are also cases in which experiments are practically impossible or infeasible, as in the assessment of human-induced ecological risks or the economic impact of a certain policy. Even in those cases where experiments are difficult or unrealistic, however, observational data of putative causes and effects may be available. For instance, collecting smoking history and health records should be much easier and ethically less problematic than running an RCT trial. Can't we make use of such observational data to say something about causal relationships? To look at it from another perspective, RCT is a method for *creating* counterfactual what-if situations through random assignments. But isn't there a way to just "peep into" other possible worlds, without creating them in this actual world?

From a metaphysical standpoint, peeping into other possible worlds may sound like crying for the moon. There is, however, a way to make this metaphysical impossibility partially possible, with the aid of a certain assumption. To see what this assumption is, let us revisit Table 5.1. Our "fundamental" problem, recall, was that although the estimation of causal effect demands us to determine both of the two rows Y_0, Y_1 for each subject, what we can observe is at most one of them. Indeed, the value of Y_0 is missing for subject A, whose X value is 1, while Y_1 is missing for B, whose X value is 0, and so on. This is a matter of course, given that the value of Y_i was defined to be "the value of Y when $X = i$." Suppose, however, that the subjects A and B are identical twins, who are almost exact replicas of each other except for their preference for sweets. In this imaginary situation, we would be able to interpret the outcome of subject B as "the outcome A would exhibit if A were not a sweets lover," and conversely, that of A as "the outcome B would exhibit if B were a sweets lover." In fact, the two need not be actual twins for the purpose of our reasoning. Regardless of their biological origin, if two or more subjects are alike in every respect except for their value of treatment X, then the value of Y_i observed for one of those subjects should be able to serve as a surrogate for "the values in the possible world" of their counterparts, which are inevitably missing in this actual world. If this is possible, by accumulating these data we can calculate the difference in effect $Y_1 - Y_0$ for each pair of replicas, and by averaging them we can eventually estimate the average treatment effect (Table 5.2).

Of course, this is a big *if*, since actual data can hardly ever contain such fortuitous twins. Besides, there is a conceptual issue regarding the ambiguity in the notion of "likeness." Since one can come up with as many possible attributes as one pleases, any pair can be considered both similar and dissimilar in infinitely

TABLE 5.2 Aggregating subjects {A, B} and {C, D} in Table 5.1 as "alike pairs." Supplementing the values of Y_0, Y_1 from the counterparts allows us to calculate the treatment effect $Y_1 - Y_0$ for each pair.

Subject	A	B	C	D	E	...
X	1	0	0	1	1	...
Y_0	0		1		−	...
Y_1	1		1		0	...
$Y_1 - Y_0$	1		0		−	...

many respects. Hence, in order to meaningfully talk about likeness, we need to delineate a privileged set of properties, such that two subjects need only coincide with respect to those properties in order for them to be judged to be alike. Indeed, requiring subjects to match in every respect is too much for our purpose. Our goal, recall, is to make X and Y_i in the formula (Equation 5.2) independent. It suffices for this purpose to find a set Z of variables for which the conditional independence $X \perp_p Y_0, Y_1 | Z$ holds. This condition is known as the assumption of a *strongly ignorable treatment assignment* or *ignorability* for short. The assumption can be written as the probabilistic formula

$$P(x| y_0, y_1, z) = P(x|z). \tag{5.3}$$

What we need, then, is a list (or vector) of attributes Z satisfying Equation (5.3). Now, to condition a probability distribution on a variable is to narrow down the subjects according to that variable, limiting our focus to the probabilities of those individuals who have the same value for that variable. The key idea in this reasoning, therefore, is that subjects with the same value of Z can be considered "alike" for the purpose of inferring a causal relationship between X and Y.

Specifically what kind of attributes, then, should be contained in Z? The answer is: all confounding factors between X and Y (Section 5.1). A Z that covers all such confounders realizes the independence condition (Equation 5.3) and allows for the estimation of the average treatment effect (Equation 5.1) from observed data. But as we noted in Section 5.1, there are in general many possible confounding factors, and including all of them in a regression model as covariates reduces the accuracy of the analysis. This problem will be mitigated if all these confounders can be summarized into one variable. There exists such a propitious variable, called the *propensity score*, which is defined as the probability of getting treatment X given covariates z, namely $P(X = 1 | z)$. In our case, two subjects can be considered alike if they coincide in just one respect, which is their probability of being a sweets lover. You may think, however, that being a sweets lover or not is hardly a conspicuous trait that one can discern with a casual look. But if you are provided with some background information, such as that a person frequently checks gourmet magazines, or has a habit of eating between meals, then you may be able to infer the probability of the person's having a sweet tooth. Thus, the strategy commonly employed in practice is to estimate the propensity score from (confounding) factors that may affect the treatment, and then substitute these estimates for z in Equation (5.3) to achieve ignorability.

The suite of causal inference techniques based on the ideas of potential outcomes and propensity scores is known under the name of the *counterfactual model*. Pioneered by Donald Rubin (1974), this methodological framework has been widely used for the task of inferring causal relationships from observed data (Morgan and Winship 2007; Rosenbaum 2017). Just like any statistical method, this approach has its own assumptions. Most important among them is the aforementioned condition of a strongly ignorable treatment assignment, the satisfaction of which requires the user to cover all possible confounders. But as we have seen in Section 5.1, there is no *a priori* limit on the number of potential confounders, and there are variables like the M-structure which should not be included as covariates. Thus the problem of covariate selection, which was the Achilles heel of multiple regression, still plagues the counterfactual model. What, then, is the merit of the counterfactual model over the conventional regression analysis method? One advantage is epistemological: by summarizing various covariates into one propensity score without explicitly including them all in the model, the counterfactual model effectively reduces the number of parameters that need to be estimated, thereby improving the performance of the model (Chapter 4) and allowing for flexible modeling. For empirical modelers who want to cover a diverse range of covariates of different nature in order to accurately estimate the causal effect under study, this is undoubtedly a big advantage of the counterfactual model. But also important, and perhaps more so for our philosophical analysis, is its semantic implications. In contrast to the regularity theory, which equates causality with some form of probabilistic relationship, the counterfactual model eschews such a reductionistic definition. It conceptualizes a causal effect not as some actual state of affairs manifest in the data, but as a sort of "inter-world difference" that emerges only when one compares observable data with counterfactual data which would have obtained under different circumstances. The concept of potential outcome, which is deeply rooted in the possible-world framework, is introduced in order to represent such counterfactual relationships. The counterfactual model and regularity theory thus part ways in their conceptual understanding of what causal relationships are. Granted, when it comes to actually estimating the average treatment effect, the counterfactual model still resorts to the traditional concept of a conditional dependence between the response and explanatory variables, which is nothing but a sort of statistical association. But such a probabilistic relationship does not *define* causality; rather, it is used as an actual-world *foothold* or *proxy* for inferring the causal relationship. The concept of causality belongs to the realm of possible worlds, which extends beyond probability models, and for this reason it is irreducible to probability. To sum up, the counterfactual model stands on the Lewisian semantics of causality, according to which causal statements are claims about the structure of possible worlds, and furnishes an epistemology for determining the truth value of such claims on the basis of observed data. In this respect, it has rich implications for metaphysical as well as epistemological inquiries into the nature of causality.

5.3 Structural Causal Models

The counterfactual model offers a powerful methodological framework for causal inference based on observational data. But just like any other method for inductive reasoning, it stands on certain presumptions about the objects of research.

We have seen some of the assumptions made by the counterfactual model, such as the inclusion of major confounders in the covariates and the exclusion of M-structures. From a mathematical perspective, the motivation and intent of these postulates is clear: they establish ignorability and allow for unbiased estimation of the average treatment effect from observed sample means. But here, we want to push a bit harder and ask what their philosophical purport is. By claiming that these assumptions make causal inferences possible, we are embracing a certain understanding about the relationship between causality and probability. What, then, is this relationship, and how can we make it more explicit?

In Chapter 1 we noted that positing a probability model amounts to assuming a uniformity of nature that remains invariant behind the randomly fluctuating data, and to build a parametric statistical model is to categorize this posited "nature" as a certain probabilistic kind. These statistical assumptions embody the ontological stance with which we conceptualize what there is and what kind of things our research targets are. Now, we observed earlier that the modeling assumptions of causal inference are different in kind from those of conventional statistical models. What kind of ontological picture, then, is implied by these assumptions? For instance, beneath the claim that covering all confounders ensures ignorability, there must be a certain assumption regarding the relationship between causality and probability. The rest of this chapter aims to extract the ontological implications of this and other modeling assumptions of causal inference, in light of yet another important methodological framework known as *structural causal models* (Spirtes, Glymour, and Scheines 1993; Pearl 2000).

5.3.1 Causal Graphs

The counterfactual approach understands causality in terms of counterfactual relationships between actual and possible worlds. But a more intuitive and common understanding of causality would take causality as a kind of directed influence, such that X causes Y if and only if X influences Y in some way or another. This kind of influence relationship is conveniently expressed by an arrow, $X \to Y$. If there are multiple causal relationships, we can add further arrows to form a directed graph to represent the causal structure over variables. Formally speaking, a directed graph is a pair (V, E) of a set V of variables and a set E of arrows between them. A directed graph used to represent a causal structure is called a *causal graph* over variables V and is denoted by G. Various causal notions can be defined in terms of graphical structures. A sequence of arrows between one variable X and another Y is called a *path*. Arrows constituting a path need not be all pointing in the same direction. But if they are uniformly directed, as in $X_1 \to X_2 \to \ldots \to X_n$, the path is called *directed*. When there is a directed path from X to Y, then Y is considered to be an effect of X. A directed path that starts and ends at the same variable, as in $X_1 \to X_2 \to \ldots \to X_1$, is called

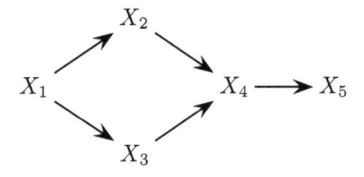

FIGURE 5.1 Illustration of a DAG. In this graph, the path $X_1 \to X_2 \to X_4 \to X_5$ is directed, while $X_1 \to X_2 \to X_4 \leftarrow X_3$ is not. X_4 is a collider in the latter path.

a *cycle*. A directed graph that does not contain any cycles is called a *directed acyclic graph*, or DAG. For the sake of simplicity, in what follows we limit our focus to DAGs. A variable on a path wedged between two incoming arrows (e.g., the Z in $X \to Z \leftarrow Y$) is called a *collider*; otherwise (i.e., if $X \to Z \to Y$ or $X \leftarrow Z \to Y$), the variable is called a *non-collider*.

The graphical representation presented in Figure 5.1 serves to capture our intuitions about causality. In particular, a graph can visually represent which variables are causally related to which and, more importantly, when such causal connections are blocked. For instance, our intuition tells us that if there is a causal "flow" $X \to Z \to Y$, then X influences Y, but also that when the intermediate Z is blocked or fixed, the "flow" is stemmed and the influence relationship is lost. This intuition about the "blocking" of a causal pathway can be refined into a slightly more formal definition. We say that a set of variables **Z** *blocks* a path between X and Y if one of the following obtains:[5]

1. There is a non-collider on the path that is contained in **Z**.
2. There is a collider on the path such that neither it nor its effects are contained in **Z**.

A non-blocked path is called *open*. The first condition should be intuitive. A non-collider on a path connecting two variables is either the intermediate of a directed path $X \to Z \to Y$ or a common cause $X \leftarrow Z \to Y$. Intuitively, if either of these is blocked, the causal connection will be cut off. The second condition is a bit complicated, but it should not require much stretch of thought to think that a path is blocked if it contains a collider (forget for now the proviso that neither it nor its effects are included in **Z**). In a colliding path $X \to Z \leftarrow Y$, X and Y independently influence the collider Z; so there is no reason to think that these causes are themselves causally related. Harder to swallow may be the proviso that this path will be open if either the collider or its effect is contained in **Z**. I just beg you to accept this as a formal definition of blocking. Or, you may imagine the causal influences as flows of water running down from the two reservoirs X and Y to Z; then a damming at Z or somewhere downstream causes overflow, giving rise to a flow between the two reservoirs X and Y.

While an open path creates a causal connection between variables, blocking it will interrupt the connection. In general, a pair of variables X, Y may have zero, one, or more paths between them. If there is no path, or if any existing path is blocked in the aforementioned sense, X and Y are said to be *d-separated* by Z. Let us illustrate this with Figure 5.1. In this figure, there are two paths between X_2 and X_3, namely $X_2 \leftarrow X_1 \rightarrow X_3$ and $X_2 \rightarrow X_4 \leftarrow X_3$. Setting $Z = \{X_1\}$ will block both, so that X_2 and X_3 are d-separated by X_1. On the other hand, the empty blocking set $Z = \emptyset$ will leave the first path unblocked, while including the effect X_5 of the collider as $Z = \{X_1, X_5\}$ will open the second path; so neither of these d-separate X_2 and X_3. We can thus conclude that the causal connection between X_2 and X_3 in this graph is interrupted if and only if X_1 is used as the blocking set.

All we have done to this point is formulate our intuitions about causal relationships in graphical terms and supply some definitions. We haven't yet said anything about how the causal structure represented in this way relates to a probability distribution over the variables. This relation is specified by the famous *causal Markov condition*, which claims that when variables are d-separated in a causal graph, i.e., when their causal connection is interrupted, they become probabilistically independent. By letting $X \perp_G Y \mid Z$ denote that variables X, Y are d-separated by Z in graph G, the condition is written as

$$X \perp_G Y|Z \Rightarrow X \perp_P Y|Z. \tag{5.4}$$

Do not confuse the two sides of the condition. The right-hand side is a claim about probabilistic independence $\perp_P$ and thus concerns a probability distribution over variables; while the d-separation relationship $\perp_G$ on the left-hand side concerns the positional relationships of variables in the causal graph (the index G makes it clear that this relation is relative to graph G). So these statements are of a different nature. Probabilistic dependence represents a co-occurrence relationship among variables, where the occurrence of one tends to coincide with that of the other. A causal graph, on the other hand, represents a causal influence among variables, where one variable influences the other along the direction of the arrows. A causal graph and a probability distribution, therefore, capture different aspects of things. The Markov condition (Equation 5.4) purports to establish a certain relationship between these two different aspects, to the effect that causally separated variables are also probabilistically independent.

But on what ground should we believe that such a Markov relationship holds? One rationale comes from taking a probability distribution as generated from an underlying causal structure, which is represented by a DAG. To explicate this argument, we first need to furnish the qualitative representation of causal relationships in terms of a DAG with a quantitative specification, by attaching a function to each variable. The function determines the value of each variable $X_i \in V$ from its direct causes and the error term U_i specific to it. Let $\mathbf{dc}_i$ stand for the values of the direct causes of variable X_i. Then the desired function has the form:

$$x_i = f_i\left(\mathbf{dc}_i, u_i\right)$$

which is defined for each variable $X_i \in V$. These equations, called *structural equations*, quantitatively specify how variables in the model respond to their causes (and the error terms). The error terms U_i are assumed to be independent and follow some probability distribution P. Substituting these distributions into the structural equations uniquely determines the joint probability distribution $P(V)$ over all variables $X_i \in V$. Thus, a DAG G, equipped with structural equations and a marginal distribution over the error terms (called "exogenous" variables), *generates* a joint distribution $P(V)$. Any distribution generated in this way is known to satisfy the Markov condition (Equation 5.4) with respect to the underlying graph, i.e., the graph that generates the distribution (Spirtes, Glymour, and Scheines 1993; Pearl 2000). Hence, if we are to think that causal structures are adequately modeled by DAGs and sets of structural equations, the Markov condition is expected to hold in general.[6]

5.3.2 Interventions and Back-Door Criteria

To summarize what we have seen so far, structural causal models represent a causal structure in terms of a directed graph, which, combined with structural equations and probabilistic inputs into the exogenous variables, generates a joint probability distribution that satisfies the Markov condition with respect to the underlying causal graph. With this in mind, we now explore the advantage of this view: how do structural causal models serve our practice of causal inference?

The first merit of the graphical representation is that it allows for a formal definition of the concept of *intervention* and the systematic prediction of its outcomes. In the counterfactual approach, a causal effect is defined as a difference between actual and possible worlds. But more commonly, causality has been understood in connection with intervention. If eating too much sweets causes a cavity, an effective intervention on the former would change the probability of the latter—or, to consider a radical measure, the enactment of a law that prohibits the manufacture, import, possession, and consumption of any sugary food would reduce tooth decay (and also delight). On the other hand, the prohibition would have no effect should the correlation between sugar consumption and cavity be a spurious one, due to some confounding factor. This suggests that if X causes Y, there is a possible intervention on X that changes the distribution of Y (Woodward 2003). Such an intervention can be

Causal Inference   163

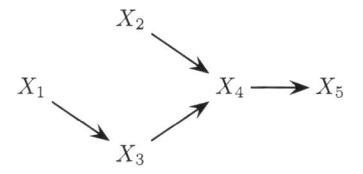

$$X_1 \to X_3 \to X_4 \to X_5, \quad X_2 \to X_4$$

FIGURE 5.2 The graph after intervention on X_2 in Figure 5.1.

expressed by deleting all incoming arrows to the target variable X in a causal graph, for the intervention should decouple it from its original causes and force it to follow some designated distribution. Figure 5.2 illustrates the graph resulting from an intervention on X_2 in Figure 5.1. If we interpret X_1 as denoting the taste of each individual, X_2 as sugary consumption, X_3 as alcohol consumption, X_4 as the presence of a cavity, and X_5 as the dental health expenditure of each person, the manipulated diagram represents that the strict enforcement of the sugar ban law brings sugary consumption down to zero, regardless of one's taste. Prohibition of alcohol, on the other hand, could be formulated as a similar manipulation on X_3 in this graph.

Next, let us ask how such interventions affect the overall probability distribution. If you are a high official of the Treasury, you might be interested in whether and to what extent the sugar ban law would reduce healthcare costs.[7] But you are certainly not supposed to actually run the experiment to see the effect. It would be beneficial if you could predict the outcome of the intervention without actually doing it, just on the basis of observational data. The *do-calculus* achieves just this, allowing for prediction of an intervention outcome using a causal graph (Pearl 2000). The calculus is based upon the causal Markov condition, which allows us to factorize a joint probability distribution into a product of conditional probabilities of the variables given their direct causes, as follows:

$$P(v) = \prod_{v_i \in v} P(v_i | \mathbf{dc}_i),$$

where $\mathbf{dc}_i$ is a value of the direct causes $\mathbf{DC}_i$ of variable V_i. In the case of the graph in Figure 5.1, the factorization yields:

$$P(x_1, x_2, x_3, x_4, x_5) = P(x_5 | x_4) P(x_4 | x_2, x_3) P(x_3 | x_1) P(x_2 | x_1) P(x_1).$$

Note that the probabilities on the right-hand side can be estimated from observational data. Now, let us intervene on X_2 to bring down the sugary consumption. We express this intervention with the operator $do(X_2 = 0)$, and write the manipulated (i.e., post-intervention) joint distribution as $P(x_1, x_3, x_4, x_5 | do$

($X_2 = 0$)). Note that in general this differs from the standard conditional probability $P(x_1, x_3, x_4, x_5 | X_2 = 0)$. To obtain this manipulated distribution, we can just apply the Markov condition again to the manipulated graph (Figure 5.2). Since the intervention deletes all the incoming arrows to X_2, we drop $P(x_2|x_1)$ from the aforementioned factorized formula and obtain

$$P(x_1, x_3, x_4, x_5 | do(X_2 = 0)) = P(x_5|x_4)P(x_4|x_2, x_3)P(x_3|x_1)P(x_1).$$

Since the terms on the right-hand side are identical to the unmanipulated conditional probabilities, this equation successfully derives the post-intervention probability (the left-hand side) from the pre-intervention probabilities. That is, given the unmanipulated causal graph and the probabilities of the variables, one can predict consequences of hypothetical interventions without actually intervening on the system. The prediction of a particular variable, say the healthcare cost after enactment of the sugar ban law $P(X_5 | do(X_2 = 0))$, can then be easily obtained via marginalization.

The interventionist conception of causality by no means conflicts with the counterfactual notion of causality discussed in the previous section; rather, they are two sides of the same coin. In effect, an intervention on variable X is nothing but a *creation* of a possible world which differs from the actual world only in the value of X, and is otherwise the same. The enforcement of the sugar ban law in the previous example, for instance, is meant to create a society devoid of any sugary foods on the market, but identical to the existing society in every other respect, including, say, alcohol consumption.[8] If, therefore, we take the difference in the expected value between two distinct intervention effects, we obtain the average treatment effect (Equation 5.1):

$$\mathbb{E}(Y_1) - \mathbb{E}(Y_0) = \mathbb{E}(Y|do(X=1)) - \mathbb{E}(Y|do(X=0)),$$

where Y_0 and Y_1 are the potential outcomes of Y under treatments $X = 0$ and $X = 1$, respectively. In other words, the average treatment effect is nothing but the difference in the expected value of Y under two point interventions, $do(X = 1)$ and $do(X = 0)$.

The second merit brought by causal graphs is that they allow us to visually inspect and identify the circumstances under which the ignorability condition, which is essential to estimating causal (average treatment) effects, obtains. In Section 5.2.2, we observed that the average treatment effect is estimable from observational data if the covariates Z satisfy the ignorability condition (Equation 5.3). But determining all the relevant covariates is, as we noted there, a tricky business. A causal graph assists us with this task by providing the *back door criterion* which graphically tells us the variables to be included in the covariates sets. A set Z of covariates, according to this criterion, achieves the ignorability condition if (Pearl 2000; Morgan and Winship 2007):

1. Z does not contain an effect of X; and,
2. In the graph where one has deleted all outgoing arrows from X, Z d-separates X and Y.

Together these conditions state that, in order to estimate the causal effect of X on Y, one should block all and only the causal paths between the two variables except for the very pathway through which X affects Y (this latter point is the reason why the second item requires deleting all outgoing arrows from X; it prevents one from blocking the very causal influence to be estimated). This criterion explains why common causes of X and Y, which create an open path between them, must be included in the covariates in order to d-separate them. On the other hand, since including a common effect $X \to E \leftarrow Y$ or a collider in an M-structure opens a path, the covariate set should not contain them. In this way, causal graphs visually guide us in determining the set of variables that should be included as covariates.

5.3.3 Causal Discovery

The discussion in the previous section presupposes that we know the true causal graph that satisfies the Markov condition with respect to the actual probability distribution. In fact, there is no reason to expect that a causal hypothesis made by a random guess would yield the correct intervention calculus or allow us to control for confounding factors. Where, then, does the correct causal graph come from? In some fortunate cases, a good grasp of the causal relationships may be provided by domain knowledge. But the causal nexus among target variables is usually unknown in many practical problems, and in such cases a causal graph must be inferred from the data in some way or another. This calls for a yet another epistemology, which probes the data for the underlying causal structure.

The methodology comes under the general rubric of *causal discovery* and comprises a variety of algorithms with different focuses. These search algorithms are motivated by the aforementioned ontological assumption, namely, that a probability distribution is generated from a certain causal structure. If so, the probability distribution and samples obtained from it should contain some signature of their origin, and by tracing this signature we should be able to recover, at least partly, the generating mechanism. Causal search algorithms trade on various sorts of such traces that persist in the data, in order to uncover a causal structure among variables in the form of a causal graph. To convey the idea, here we briefly review just one of the most standard and classical approaches that make use of statistical independence. Earlier we used the Markov condition to derive probabilistic conclusions from a given causal relationship. Relevant to causal search is the inference in the opposite direction, which infers a causal structure from the given information about probabilistic relationships. For this purpose, we assume the following *faithfulness condition*:

$$X \perp_G Y | Z \Leftarrow X \perp_P Y | Z. \tag{5.5}$$

Conversely to the Markov condition, the faithfulness condition claims that if conditional independence obtains among the variables in question (the right-hand side), then they are d-separated (the left-hand side). Let us see how this allows for causal inference. Suppose we have a probability distribution over three variables X, Y, Z for which $X \perp_P Y$ and $X \not\perp_P Y | Z$ hold.[9] That is, X is independent from Y unconditionally but becomes dependent if conditioned on Z. If faithfulness holds, there is only one causal relationship consistent with this pattern of independence, namely $X \rightarrow Z \leftarrow Y$. This is because, on the one hand, $X \perp_G Y$ guarantees that there is no directed path or common cause that exchanges causal influence between X and Y, and on the other hand, from $X \not\perp_G Y | Z$, i.e., the fact that Z opens a path between X and Y, we can conclude that Z is a collider between the two. This illustrates how the faithfulness condition allows for causal inference from probabilistic relationships.

However, full recovery of the causal structure is not always possible. To see this, suppose that the independence relationships $X \not\perp_P Y$ and $X \perp_P Y | Z$ hold. This distribution is consistent with any of the following three different causal hypotheses: $X \rightarrow Z \rightarrow Y$, $X \leftarrow Z \leftarrow Y$, and $X \leftarrow Z \rightarrow Y$. Therefore, one cannot uniquely determine which of these represents the true data-generating structure. In this and many other cases, what we can infer from a probability distribution is a set of causal hypotheses that equally accommodate the given data, and in such cases we must resort to other means in order to further narrow down the candidate hypotheses. Moreover, the assumed faithfulness condition does not necessarily hold all the time. Faithfulness means that probabilistically independent variables are causally separated. But in case one variable influences another through two distinct pathways that cancel each other out, there may be no visible statistical association between the two variables, even though the first should in fact be considered a cause of the other. When we have such an *unfaithful* distribution, the independence-based algorithm for causal discovery sketched earlier does not work out.

We thus come down to the same moral we have encountered again and again in this book: one cannot make an inductive inference without making an assumption. Probing the data for a probability distribution requires an assumption like the IID condition or a statistical model. Using a propensity score to estimate the average treatment effect presupposes the ignorability condition. Likewise, the algorithmic reconstruction of causal graphs from conditional independence calls for faithfulness or similar assumptions. These assumptions themselves cannot be justified by the statistical methods they support, and so they must be justified by other means or background knowledge, unless they are accepted as dogma. Statistics as an epistemology is a never-ending process of justification, which aims to build a bridge over the fundamental logical gap inherent in any inductive inference. The faithfulness condition is one such bridge, such that by crossing it one sets foot in the realm of causality with the aid of probabilistic knowledge. (The Markov condition, on the other hand, is a bridge going in the opposite direction, which one may use to derive a probabilistic prediction on the basis of causal knowledge.) The independence-based algorithms sketched in this section are by no means the only bridge between causality and probability: there are several other approaches which make use of, say, distributional or functional forms. For the details of such approaches, see Peters, Janzing, and Schölkopf (2017).

5.4 Philosophical Implications of Statistical Causal Inference

This chapter sketched the two major approaches of statistical causal inference, the counterfactual model and causal structural model, and examined their philosophical implications. In this concluding section, we would like to see how all these pieces fit together with the statistical worldview we have tried to portray throughout this book.

Let us begin by recalling the dualism of data and probability models that we introduced in Chapter 1 as the ontological framework of inferential statistics. In order to infer the unobserved from the observed, inferential statistics posits a latent entity beyond the data. This entity, mathematically expressed as a probability model, is expected to serve as the "uniformity of nature" that remains invariant over samplings at different times and places, and predictions are made via inference to this probability model. It is under this dualistic ontology that modern statistics has developed its sophisticated machinery for predicting unobserved phenomena and evaluating risks under uncertainty.

Does the same framework also accommodate causal inference, another typical kind of inductive inference? Until around the mid-20th century, many philosophers believed that there is no fundamental gap between causal explanation and prediction, and they aimed to locate them within a unified formal framework (the so-called *covering law model*; Hempel and Oppenheim 1948). In parallel, statisticians developed various methods for gauging causal relationships in probabilistic terms, like multiple regression, generalized linear models, structural equation modeling (SEM), and so forth. These reductionist approaches to causal inference, however, face the conceptual problem noted in Section 5.1, precisely because causal relationships are *not* probabilistic relationships, as philosophers as well as statisticians began to realize toward the end of the century. The nomological pattern underlying causal inference or everyday causal talk is different in kind from the uniformity introduced for the purpose of prediction. For this reason, probability models, or any model of the uniformity of nature for that matter, fail to capture causality.

This becomes even more evident if we consider intervention, a notion that is intimately related to our conception of causality. While prediction is an inference to a potential or future observation, causal inference concerns the effect of a hypothetical intervention (Section 5.3.2). An intervention alters the target system by interfering with its old state to turn it into a new state. This breaks down the uniformity or probability distribution of the original system and replaces it with a new one. The enactment of a sugar ban law will change the proportion of cavities as well as the distribution of healthcare costs (which is exactly the aim of the law). The interest of causal inference lies precisely in this transition and the resulting distribution. The laws of this change cannot be sought only within the very system subject to the change. What is needed for such an inference is an inter-world nomological relationship that connects different distributions/worlds, and in this sense causal inference calls for a methodological framework that goes beyond probability models.

The notions of possible worlds and causal models introduced in this chapter are the conceptual machinery we can use for capturing this realm lying beyond the probability models. Just as inferential statistics introduces probability models as the uniformity of nature in order to infer unknown data from known data, causal inference introduces causal models as another kind of law in order to derive the post-intervention probability model from the pre-intervention one. One can thus conceive them as yet another structure lying behind probability models, and with this additional ontological posit, causal inference comes to espouse a *trialist ontology* (Figure 5.3) which accommodates three sorts of entities—data, probability models, and causal models. A probability model, as we saw in Chapter 1, models the world itself, which lies behind observable data. Different probability distributions thus represent different worlds. The goal of traditional statistics discussed in the previous chapters is to help us determine, on the basis of observational data, which among these possible worlds/distributions is this actual world. On the other hand, the question to be answered by causal inference is: if one were to make an intervention on the actual world, which world would be realized? Which world would it evolve into? An intervention is a mapping from the actual to a possible world, and the goal of causal inference is to identify the law that governs this mapping. Causal models are models of such mapping relationships among possible worlds. In the newly introduced causal layer, interventions are defined as graph transformations, on the basis of which the *do*-calculus derives the post-intervention probability distribution. As a matter of fact, a causal model can be thought of as a function that takes a distribution and an intervention as inputs and gives the post-intervention distribution as output, while the *do*-calculus is a way of calculating this function. An inference to a causal effect amounts to predicting, with the aid of such

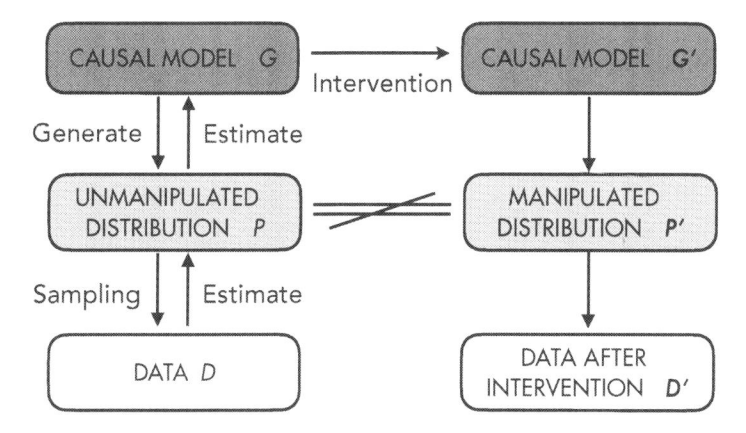

FIGURE 5.3 The trialism of causality, probability, and data. The causal inference framework posits causal models as an additional sort of "entity" behind probability models, a mechanism that generates the latter. Since an intervention breaks the uniformity of a probability model, the post-intervention distribution cannot be calculated from the pre-intervention distribution alone. Defining interventions in the causal layer allows one to make an inference to its possible consequences on the distribution as well as the data to be observed. Meanwhile, the estimation of causal models tend to be more difficult, for they lie one step "further" away from the data than probability models.

calculation, the change from the actual to a possible world induced by an intervention. And, as we have argued, this calculation becomes possible only on the supposition of the generative law behind possible and actual worlds/distributions, as well as the formal modeling of such a third-level "entity" in terms of causal models.

Clearly distinguishing these ontological layers proves important not just philosophically, but also for a better understanding of statistical concepts. In inferential statistics, it is crucial to distinguish between properties of data described by various sample statistics on the one hand, and those of the population described by, say, expected values, on the other. Though parameters of a distribution may be estimated from data, they cannot be calculated from them, for these two concepts, parameters and data, belong to different ontological layers. Exactly for the same reason, concepts belonging to probability models and those belonging to causal models must be sharply distinguished in causal inference (Pearl 2000). While expected values and independence are properties of a probability model, the average treatment effect and d-separation belong to the realm of causal modeling. They are conceptually distinct, and the one cannot be identified with the other. What we can do at best is to *infer* the latter from the former under certain conditions, such as ignorability or faithfulness. Likewise, causal graphs should *not* be identified with *Bayesian networks*, understood as a graphical encoding of patterns of conditional independence among variables. For while Bayes nets are just graphical representations of properties of individual probability distributions, the primary aim of causal graphs is to capture relationships *among* distributions, as we have stressed. Mathematical formulations sometimes hide this and other important ontological distinctions under their abstract guise. Mathematically speaking, Bayes nets and causal graphs are the same kind of object—directed graphs—in graph theory. Furthermore, the estimation of causal effects based on propensity scores uses essentially the same method as the standard regression analysis. Such abstractions exemplify the strength of mathematical formulations, which are applicable to a wide range of phenomena regardless of their specific peculiarities. But exactly for this reason, the same mathematical method or expression may stand for different things. This difference is often attributed to a difference in "interpretation"—one might say, for example, that a given statistic is interpreted as a correlation coefficient in one context and as a causal effect in another. Looking at it more deeply, however, this in fact is a difference in the underlying ontological assumption, i.e., in how we conceptualize the target phenomena and what kinds of things we think they are. If the problem under consideration is conceptualized in terms of a probabilistic kind, then all what we can conclude about it will be restricted to probabilistic matters, and the validity of our conclusions will depend on whether the target system really has the ontological features of the probabilistic kind (such as uniformity). If, on the other hand, we want to draw causal claims, we need to regard our target not as a mere probabilistic kind, but rather as a *causal kind*. Then the validity of those claims will be contingent on the soundness of such an ontological attribution, i.e., whether the conditions to be satisfied by a causal kind are really satisfied in the case at hand. For these reasons, understanding concepts at the proper ontological level is essential even in the practice of statistical analyses.

To sum up, what we can conclude with the mathematical machinery of statistics depends on the ontology with which we conceptualize the phenomena under question. Statistics encodes such ontological assumptions with the formal apparatus of probability distributions, potential outcomes, causal graphs, and so forth, and provides epistemological methods for checking these assumptions on the basis of data. On the other hand, the question of which ontological stance should be taken or preferred in the first place is not a question that admits a general or logical solution; rather, it must be determined case-by-case through considerations of the nature of the problem and the aim of the research. If our interest lies in prediction, probabilistic kinds will do the job; but if our task involves intervention or control, causal assumptions are in order. We thus have to decide on our ontological stance in accordance with our problem and aim, prior to choosing and applying a particular epistemological method suitable for that purpose.

Further Reading

Mumford and Anjum (2014) and Kutach (2014) are short introductory textbooks on the philosophy of causation. Rosenbaum (2017) explains the elements of the counterfactual model with a variety of examples and little math, while Morgan and Winship (2007) offer a more detailed account. Structural causal models are succinctly summarized in Pearl, Glymour, and Jewell (2016). Peters, Janzing, and Schölkopf (2017) is a more advanced exposition that covers various approaches to causal discovery.

Notes

1. Interestingly, Hume also proposed a counterfactual definition of causation right after his regularity-theory-like proposal, as a paraphrase of the latter (Hume 1748, sec. 7, pt. 2). Despite his equivocation, however, these two definitions are different in nature.
2. Of course, the counterfactual theory is not flawless and has faced several objections, including most notably the problem of overdetermination (Kutach 20140; Mumford and Anjum 2014).
3. "An inference from effects to causes" is the characterization that has been more commonly associated with Bayesian inferences, which can certainly be used to infer from the occurrence of the effect the probability of the cause/hypothesis, which is itself expressed as a random variable. Such an inference to a probability is barred in frequentism, where a hypothesis is not a random variable. Yet one may still take it as a reliability-based inference to the cause/hypothesis.
4. In general, if two random variables X and Y are independent, i.e., $P(Y|X) = P(Y)$, the same relation holds for their expectations: $\mathbb{E}(Y|X) = \mathbb{E}(Y)$ (verify this from the definition of expectation). Furthermore, since in the present case Y_i is a binary variable that takes either 0 or 1 as its value, its expectation corresponds to its probability, i.e., $\mathbb{E}(Y_i) = P(Y_i)$.
5. Here $\mathbf{Z}$ and $X \cup Y$ are assumed to be mutually exclusive.
6. Conversely, one who thinks that there is something in causality that eludes this modeling machinery is well motivated to doubt the causal Markov condition. See Cartwright (1999) for a skeptical view.
7. Note that this book was originally published in a country that has a national healthcare system.
8. An intervention that does not satisfy this condition and affects variables other than the target is called *fat-hand*.
9. Of course, as we have repeatedly emphasized throughout this book, a probability distribution is not something that is "given," but must be inferred from data. More specifically, whether the said independence relationship holds or not must be judged by using some statistical procedure like hypothesis testing. But in what follows, we will simply take for granted the correctness of such judgments, and instead focus on the inference to a causal hypothesis given knowledge about a probability distribution.

6
THE ONTOLOGY, SEMANTICS, AND EPISTEMOLOGY OF STATISTICS

In this book we have explored various methodologies in statistics, including Bayesian as well as classical statistics, model selection, machine learning, and causal inference, from the philosophical perspectives of ontology, semantics, and epistemology. In this concluding chapter, we summarize the discussions and reflect on how each of these philosophical threads comes together to weave the fabric of statistical thinking.

6.1 The Ontology of Statistics

Any empirical science comes with its own ontological assumptions specifying the objects of its investigations, as well as ontological posits necessary for its explanatory practices. Based on this observation, we began this book by taking an ontological inventory of statistical science. What kind of things must be given or assumed in inductive inference? The ontology of statistics concerns such ontological questions and tries to identify the basic building blocks to be used in statistical inference and explanation.

The most obvious and fundamental "thing" in statistical practice is data. No data in, no statistical inference out. Descriptive statistics begins with data and aims to extract characteristics and patterns from them in the explicit form of sample statistics. In this way, it contributes to our economy of thought by organizing a disorderly jumble of numbers into comprehensible figures. On the other hand, descriptive statistics remains within the strict boundaries of the observed and says nothing beyond that, including what unobserved phenomena will look like. Such an inductive inference calls for an additional entity that serves as the latent source of the data, or what Hume called the uniformity of nature. A probability model is a mathematical model of the structure of the world, which is supposed to remain invariant over observations at different times and places. Standing upon this data-probability dualism and taking the former as partial samples from the latter, inferential statistics offers a mathematical framework for estimating the latent model from the data and using it to predict unobserved phenomena. Since we can never directly access the probability model, the assumption of uniformity must always remain a yet-to-be-confirmed hypothesis. Nevertheless, we cannot do without hypothesizing such an latent entity if we are to predict the future on the basis of past experience.

If we take a probability model to be a model of the way the world is, then prediction can be thought of as an inductive inference bound within a single fixed world. This is in contrast to causal inference that we discussed in Chapter 5, whose main interest is the outcomes that would ensue when one (hypothetically) makes changes to the world. That is, causal inference is interested in whether and how a given intervention changes a probability distribution. A causal statement that a variable X causes another Y is tantamount to the counterfactual claim to the effect that if X's value were different, or if it were manipulated, the distribution of Y would be different. In this sense, the objective of causal inference is to understand inter-world relationships, or, more accurately, transitions from one possible world to another triggered by interventions. Such inferences cannot be bound within a single world or probability model; one needs to further assume a plurality of possible worlds, connected to each other via certain nomological relationships. What is represented by causal models are such relationships among multiple possible worlds or probability models. There, interventions are defined as manipulations of causal graphs, on the basis of which a set of probability distributions is mapped to a set of post-intervention distributions. A causal statement that one variable is the cause of another is understood as a statement about the law governing the inter-world mappings. Not only is such a law invisible in the data, it also cannot be captured by a single probability distribution *qua* model of the actual world. In this sense, causal inference stands upon a deeper ontological assumption and takes on the difficult task of inferring this inter-possible-world structure from data obtained in the actual world.

Ontological assumptions, therefore, determine the type of explanations available in a given statistical theory. While the positivist data-monism will suffice for the economy of thought, the dualistic ontology of probability models and data is in order for making predictions. And causal explanations and the intervention calculus call for yet another, deeper, ontological layer of causal models. In general, a richer ontology enables a wider range of inferences. However, it also places a higher epistemological burden on us, for a heavy ontology means that more must be identified from the data. Empirical investigations always proceed from a more direct and superficial layer to deeper and more obscure layers. We set out with the data at hand and use them to probe a probability model, which in turn is used to probe a causal model. The validity of each inferential step depends not only on the methodological assumption proper to that particular level, such as the IID assumption or the faithfulness condition, but also on the correctness of the prior inferential steps. A correct causal inference thus assumes not only the ignorability or faithfulness condition to bridge the gap between the layers of probability and causality, but also that the probability distribution that serves as the input for the causal inference is correctly estimated, which in turn hinges upon various conditions that bridge the layers of data and probability. Since there is no foolproof guarantee that these assumptions are true, inferences to a deeper layer tend to be challenging and uncertain. Given the inevitability of this trade-off between explanatory power and epistemic burden, it is crucial in the practice of statistical inference to make just the right kind of ontological assumptions, in accordance with the nature of the desired explanation or inference.

Making clear distinctions among ontological layers proves useful in understanding various statistical concepts and their estimation procedures, too. Although most concepts in statistics are defined as a quantity or function, this does not mean that they all have the same ontological status—rather, as we have seen in Chapters 1 and 5, different concepts live in different worlds, as it were. While sample statistics live in the realm of data, concepts like expected values, probability distributions, parameters of a distributional family, and coefficients of a regression model belong to the realm of probability models. Finally, the average treatment effects and coefficients in structural equations pertain to causal models and represent inter-possible-world relationships. Estimation is the art of capturing concepts lying at a deeper layer using those at a more superficial layer. It includes estimating parameters or regression coefficients from appropriate sample statistics within certain error bounds, evaluating causal effects via estimated expected values, and judging the presence or absence of a causal connection between variables on the basis of statistical tests. Statistics can thus be conceived as an attempt to cross these ontological boundaries, as well as determine the conditions under which this is possible.

Since any layer beyond the data is not open to direct access, such a "transgression" inevitably faces serious epistemological challenges. To overcome this difficulty, standard statistical practice bases its inferences on further simplifying assumptions that curtail the hidden entity into prespecified types. Statistical models, or probabilistic kinds in our parlance, are such types (Chapter 1). They are, so to speak, "models of a probability model"—the result of carving a probability distribution, itself something amorphous, into functions having definite analytical forms. Such an explicit formulation enables us to represent a probability distribution using a finite number of parameters, and to categorize individual stochastic phenomena into well-defined kinds or types of inductive problems, each represented by a distributional family such as a Bernoulli or normal distribution. In this way, probabilistic kinds sort out *sui generis* probability models and organize them into universal archetypes, just as natural kinds like "platinum" and "tiger" carve up nature and phenomena into handy categories useful for prediction and inference. We believe that these natural kinds that we use in science as well as in everyday life reflect the objective structure of the world and "carve nature at its joints." But we can never know for certain whether our way of carving really follows nature's joints. Just as what had been called jade turned out to be a category encompassing two sorts of minerals with different chemical constitutions, nephrite and jadeite, what we accept as a natural kind today may later be discovered to be an incorrect demarcation of nature. Likewise, a probabilistic kind represents *a*, but not *the*, way of carving nature. It is only one among many possible ontological hypotheses that helps us understand a probability model and compare numerically distinct stochastic phenomena.

Another point we should consider in connection with the ontological status of probabilistic kinds is the pragmatic aspect: what do we expect of such natural kinds, and for what purpose do we "carve nature" in the first place? If we expect natural kinds to reflect the objective structure of the world, good probabilistic kinds should be those that faithfully copy the target probability model. There is, however, another perspective, according to which the role of natural kinds is not so much to trace nature's joints as to structure our experience in such a way as to facilitate future predictions, or in other words, to identify the real patterns (Chapter 4) that robustly show up in past and future observations.

These two conceptions represent different ontological stances regarding the nature as well as the role of natural kinds. In Chapter 4, we described such an ontological shift as being prompted by theories of model selection. In contrast to traditional statistics, which tries to close in on the data-generating process in terms of probabilistic kinds, model selection theories shift the goal and aim to find a statistical model that will accord well with potential sample distributions to be obtained in future observations. These two aims do not necessarily concur, because the nature of the data we will observe is conditioned by our epistemic capabilities and depends on pragmatic factors such as sample size. Which patterns are considered "real" usually depends on the cognizer's epistemic abilities. Much in the same way, which probabilistic kinds are to be carved out by model selection criteria as robust patterns that are useful for prediction depends not just on nature, but also on the pragmatic circumstances of whoever uses them for predictive purposes.

Dennettian pragmatism thus suggests that our ontology should be tailored to our epistemic capabilities. This also implies that a better cognizer equipped with bigger data and more powerful computational abilities would be able to discern real patterns in what only appears to us as senseless noise and use them for predictive purposes. This seems to be realized by the recent developments in machine learning, which offer us glimpses into machine-cognizers that partly surpass our cognitive capabilities. What is interesting about deep models is not only that they are themselves gigantic probabilistic kinds with a huge number of parameters, but also that they seem to represent and classify the given data—and thereby carve up the world—in their own way. This naturally leads to the question as to what *their* ontology looks like. Do deep learning models carve nature as we do, or are they using completely different natural kinds? The advance in transfer learning, which shows that models trained in one domain can be successfully applied to a different domain, seems to suggest that natural kinds learned by deep learning models are somewhat carved out according to nature's joints. On the other hand, the striking cases of adversarial examples (Section 4.4.3) suggest the possibility that their carving is not yet perfect, or at least that the kinds they identify as "natural" may be utterly different from ours. Since these problems may present serious potential risks in the social application of deep neural nets, consideration of their ontology not just is of a philosophical interest, but also concerns the evaluation of the deep learning technique in a broader context. For one thing, ontology is at heart of our understanding of others. We cannot communicate with and understand those who have an utterly different ontology from us. Even our predictions and anticipation of behaviors shown by other people or animals depend on our understanding of their ontology. For this reason, the application and acceptance of the deep learning technique in our society must be accompanied by an elucidation of its ontology. This is so even if it is an intrinsically hard problem with no single correct answer, as we suggested in Chapter 4.

6.2 The Semantics of Statistics

While ontology stipulates the type of "entities" to be assumed in statistical inferences, semantics concerns how these mathematical posits relate to the actual world. From a broader philosophical context, this pertains to the problem of interpretation or representation of theoretical models. Like the empirical sciences, statistics builds models of stochastic phenomena in order to study their properties and behavior. But since such models are idealized abstract entities distinct from the targets themselves, the use of a model naturally raises the question as to what, if anything, in the real world do the model itself and the results derived from it correspond to. This sets the question to be answered by the semantics of statistics.

Semantic questions arise at each ontological layer discussed previously. The most intensively discussed among these is the semantics of probability models, that is, the problems concerning the nature and meaning of events and probabilities. As is well known, this question has yielded two conflicting answers, subjectivism and frequentism, and has provoked fierce debates between them. According to the subjectivist interpretation canonical in Bayesianism, an event stands for a proposition, and a probability represents the degree of belief in that proposition. In contrast, frequentism defines an event as an objective state of affairs, and probabilities as its limiting frequency. While we did not go into the fine details of these semantic claims in this book, rigorously establishing these corresponding relations in general will call for a certain *representation theorem*. Given a probability model on the one hand and some system of the actual world (which, for instance, would be a set of propositions ordered by the preference of an epistemic agent in the case of subjectivism, or a set or *collective* of events and their relative frequencies in the case of frequentism) on the other, the desired representation theorem would secure the correspondence between these two structures by establishing a homomorphic relationship between them. The homomorphism will then guarantee that one can coherently translate statements about the probability model into statements about corresponding aspects of the world. From this it should be clear that the problem of probability semantics, by its nature, concerns the interface between mathematical models and the actual world, and not the mathematical study of formal models themselves—*a fortiori*, semantics does not license, guide, or constrain particular mathematical derivations or proofs. In this sense, statistics, as a branch of applied mathematics, may and does proceed independently of its semantic characterization, just as the ongoing research in quantum mechanics does not depend on its interpretation (say the Copenhagen interpretation), and number theorists can get on with their theorem-proving work without being concerned about the applicability of numbers to real phenomena (e.g., Krantz et al. 1971). Indeed, the biggest advantage of building mathematical models is that it enables us to study empirical phenomena as if one were dealing with a problem of pure math, without worrying about interpretative issues. However, it is ultimately the job of semantics to guarantee that this kind of mathematical investigation also represents a study into *nature*.

In addition to providing an intelligible interpretation by which we can anchor statistical formulas in the real world, semantic analysis plays the further, negative but important role of preventing meaningless inferences and conclusions. The analysis of the conditions of meaningfulness—which may be called a "critique" in the Kantian or Wittgensteinian sense—has been the major motivation behind the *theory of measurement*, which studies the conditions of the applicability of mathematical quantification and operations to empirical phenomena (Narens 2007). In statistics, one often encounters this problem of meaningfulness in the form of the distinction between various types of scales of measurement. It does not make sense, for instance, to compare averages of ordinal variables, because they are not invariant under transformations that preserve order. That is, such averages can change even if the ordering as a whole remains the same, if we were to assign different numbers to the places in the ordering. This means that the "average of an order," although calculable, is not a meaningful concept. The criterion of meaningfulness and the range of mathematical operations meaningfully applicable to a given variable depend on what the variable represents. This is also true for probability values. That is, what we can meaningfully discuss about probabilities is determined by what they represent. As we have seen, if we adopt the frequentist interpretation, which equates probabilities with limiting relative frequencies of events, we cannot meaningfully talk about such a thing as the "probability of a hypothesis." The subjectivist interpretation, on the other hand, does license the assignment of probabilities to hypotheses, but these probabilities are degrees of beliefs of individual epistemic agents and do not measure the objective correctness of the hypotheses. In this way, one's choice of semantics circumscribes what one can meaningfully say and conclude about probability values.

Semantic questions also arise in connection with causal models, which reside in a one-step-deeper ontological layer. Indeed, now that the battle over the semantics of probability waged between subjectivism and frequentism has largely been settled toward the turn of the century, the semantics of causality is emerging as a more relevant topic. Despite being an extremely common term, causality has long been a concept shrouded in mystery. Ever since Hume, philosophers have tried to understand what is and should be meant by a causal claim, such as that X causes Y. The two approaches to causal modeling discussed in Chapter 5 propose answers to this semantic question, albeit in slightly different ways. Counterfactual models introduce potential variables that represent possible worlds and define a causal effect as a difference in their values between multiple potential/possible worlds. Structural causal models, on the other hand, define interventions on graphs and identify a causal effect as a change in the distribution of an effect variable induced by an intervention. These explicit definitions of causal relationships in terms of mathematical frameworks open up the possibility of empirically estimating causal effects. At the same time, the semantics delineates the range of empirically meaningful questions about causality. For instance, can race or sex be a cause of other variables (Marcellesi 2013; Glymour and Glymour 2014; VanderWeele and Robinson 2014)? That is, can one meaningfully consider interventions on these attributes, which are often seen as essential constituents of personal identity? If the answer is negative, a causal claim about the influence of gender on salary, say, will be regarded as meaningless under the interventionist interpretation of causation, and its truth value will be indeterminable.

Semantic analysis, therefore, precedes epistemological investigation and lays down the conditions of its possibility. Once a certain semantic framework is set, statistics *qua* epistemological practice can be carried through as a mathematical study of formal models, without worrying much about whether it corresponds with the real world. For this reason, the importance of semantics is not conspicuous in well-developed scientific disciplines. Physicists and psychologists may take for granted that the phenomena they study can be represented by and investigated within certain mathematical frameworks, and hence feel little need or interest in justifying their semantic assumptions. Likewise, some statisticians may regard interpretative questions as a relic of the past, or even a hindrance to the steady development of mathematical statistics. In one sense, such a skepticism may be a healthy sign of the maturity and soundness of the discipline. Be that as it may, semantic questions will never disappear. This is because, while statistics has a highly developed mathematical system, it cannot be confined within mathematics. Indeed, statisticians build formal models, calculate probabilities, and derive logical consequences; but what's more, these conclusions must be interpreted in the context of concrete phenomena if one wants to solve real problems. What is meant by a low or high p-value, AIC score, or posterior probability of a given hypothesis? Unless given a concrete meaning, these values do not appeal to people's minds or affect the social attitude toward the hypothesis in question. Statistics is a mathematical science that studies the deductive consequences of formal models, and at the same time it is an empirical science that aims to apply these models and conclusions to real-world problems. Because of this Janus-like nature, it cannot cast off the semantic issues of how its mathematical machinery comes to have an empirical significance.

6.3 The Epistemology of Statistics

Once posited and interpreted, probability models can be put in service of carrying out inductive inferences from data. In this book we have used the term "epistemology" to denote the core part of statistics, which deals with this inferential procedure. Statistical inference carries within itself an essential difficulty, in that it tries to fathom the ungiven from the given. The ultimate goals it aims for are sometimes infeasible, as in the case of knowing the nature of an entire population, and at other times a sheer metaphysical impossibility, as in the case of the fundamental problem of causal inference (Chapter 5). We cannot, therefore, unconditionally and uncritically accept conclusions drawn from statistical procedures as knowledge in their own right. Nonetheless, consequences of a sound inference may well be regarded as justified, and statistics does assume a role of justifying scientific hypotheses in modern society. From this follows the epistemological question regarding in what sense conclusions of statistical inferences are justified. When we make predictions or judge hypotheses under the guide of statistics, on what grounds can we be sure that we are making valid decisions?

Driven by this question, we have characterized the various methodologies of statistics as epistemologies standing upon different concepts of justification. Bayesian statistics is comparable to internalist epistemology, with its emphasis on the logical coherence among beliefs, while classical statistics is comparable to externalist epistemology, which places emphasis on the reliability of the belief-acquisition process. By no means is this comparison intended to suggest that there is a perfect analogy between the statistical methodologies on the one hand and the philosophical doctrines on the other. The simple dichotomy proposed in this book hardly does justice to the nuanced and complicated practice of statistics and epistemology in the literature. Just as all statistical models are wrong, our meta-statistical analysis distorts and deforms the true nature of both disciplines. We believe, however, that drawing a connection between the two distinct scholarly traditions in terms of a simple and idealized model proves "useful" not only for characterizing their epistemic natures, but also for elucidating the fundamental challenges they confront in the face of the problem of inductive inference.[1] One among several perspectives offered by this parallelism is the reinterpretation of the partisan debate between the Bayesian and frequentist schools as an epistemological controversy over the "correct" concept of justification. At the heart of this controversy is the truth-conducive property of justification, one of the central topics of the modern epistemology. Bayesians evaluate the degree of belief in a hypothesis using the internalist inferential calculus, in such a way as to cohere with the data and likelihood. But how and why does this kind of internal coherence warrant truth, understood as a correspondence with the outer world? This is a fundamental question for internalist epistemology in general, and to tackle this issue, internalists need to step out of their belief system and take external factors into account (Section 2.3.3). On the other hand, in classical statistics the decision of rejecting or not rejecting a hypothesis is justified on the basis of the reliability of the testing process. Such an externalist justification, however, is truth-conducive only when the process is actually trustworthy and properly employed. Since a test's verdict or p-value remains silent about the validity of these external conditions, the proper use of a statistical test calls for an independent check on the reliability of its justificatory machinery (Section 3.3.3).

The holy grail of traditional epistemology—both internalist and externalist—is the truth. In the context of statistics, this amounts to seeking for a hypothesis that faithfully captures the probability model, and the justificatory concepts of traditional statistics, both Bayesian and classical, aim to warrant this objective correspondence. The truth, however, is not necessarily the sole aim of statistical practice. The primary motivation of introducing statistical models, we recall, is the prediction of unknown phenomena; it might make more sense, then, to evaluate statistical hypotheses on the basis of their predictive performance. This idea led us to a kind of epistemic pragmatism, which puts more epistemological value on the performance or utility of a statistical hypothesis than on its veracity. Chapter 4 introduced the theory of model selection and the deep learning technique as major approaches along this line of thought. The theory of model selection offers a theoretical framework and criteria for assessing the generalization capability of models, while the rapid development in the machine learning literature has enabled us to build and train models whose abilities far surpass those of conventional statistical models. In these engineering pursuits, the epistemological excellence of a model is only defined relative to the nature of a given task, and it is effectively determined by the loss function used to evaluate the model.

However, this does not necessarily constitute a vindication of a Quinean "naturalized epistemology," where philosophical epistemology loses its independent *a priori* footing and is absorbed into the edifice of scientific research, perhaps as a sort of cognitive psychology (Quine 1986). The normative question about the nature of justification will retain its significance, especially when these model evaluation methods are applied to scientific investigations. This is particularly pertinent to the increasing applications of deep learning methods in scientific contexts: in what sense do they justify hypotheses, and what should we make of scientific discoveries made by machines? Adopting machine-aided reasoning as a new method of scientific investigation, and its results as genuine knowledge, will have serious ramifications for our traditional conception of science. Ever since the time of Galileo, modern science has developed under a foundationalist ideal. This is most famously vindicated by the father of modern epistemology, Descartes, who argued that scientific theories must be built upon a solid and clear foundation through the careful and conscious stacking of valid inferential steps. This kind of foundationalist ideal of science still has a strong appeal today. In statistics, too, Bayesianism and classical statistics show a similar foundationalist tendency in their attempt to ground their conclusions on explicitly stated theories and principles. In contrast, deep learning models lack a unified theoretical underpinning that can be used to quality-control their reliability, which therefore must be assessed by *a posteriori* experiments. In the absence of an *a priori* theory of justification, should we count their findings as scientific knowledge? Chapter 4 takes up the perspective of virtue epistemology as a clue for approaching this question. Virtue epistemology seeks the ground of justification not in a universal theory but in the personal capability of an individual epistemic agent. Likewise, the contemporary machine learning literature seems to justify conclusions of deep learning models on the basis of their individual or "personal" characteristics, such as benchmark scores, model architectures, and even the model's creators. If so, the validity or truth-conduciveness of such justificatory procedures should hinge upon the nature of the alleged "epistemic virtues," and most importantly, whether they really constitute "virtues." And from a broader context, there is also the question as to whether we are willing to accept the findings justified in such a way as scientific knowledge. For one thing, the core tenet of virtue epistemology—to seek the basis of knowledge in the virtues of individual models or persons—may somewhat appear to be a regression to obscurantism, especially in light of the modern scientific ideal that puts objective and universal laws as first principles. This, or perhaps a similar impression, may be what lies at the bottom of the anxiety people feel toward the ongoing technological revolutions surrounding deep learning. If so, an epistemological inquiry into statistics should still be of great significance today.

6.4 In Lieu of a Conclusion

Guided by these ideas, in this book we have explored how contemporary statistics takes on the challenge of inductive inference, the long-standing philosophical conundrum since the time of Hume. In particular, in various parts of the book I have tried to draw a parallelism between statistics and philosophical epistemology. I hope this analogy helps to bridge the heretofore-unrelated two research traditions and to shed light on the problems characteristic of each. But readers are again reminded that this is just one perspective or model. To what extent this model is faithful to actual practice or useful in understanding the theories of both statistics and epistemology, I defer to the reader's judgment. I have also tried in this book to cover some new topics that have not been discussed much in the traditional philosophy of statistics, such as deep learning and causal inference. On the flip side, the exposition and discussion of each topic has admittedly turned out terse and sometimes partial. Furthermore, I have not been able to address some of the major statistical topics like Bayesian hierarchical modeling, interval estimation, kernel methods, and recent developments based on information theory, despite their evident conceptual as well as practical importance. On the philosophical side, our coverage of epistemology has been limited to topics mainly developed in the last century, and leaves out all the recent trends in, say, contextualism, inferentialism, and social epistemology. How these topics can be analyzed from a philosophical or statistical perspective is an open question, for which I invite the reader to join in on the inquiry. If the reader, after finishing this book, finds such a challenge appealing and worth pondering, I take it that the book has well achieved its initial promise of introducing philosophy to data scientists, and data science to philosophers.

Note

1. In this respect, we concur with Wimsatt's (2007) slogan that false models are "means to truer theories."

BIBLIOGRAPHY

Adadi, Amina, and Mohammed Berrada. 2018. "Peeking Inside the Black-Box: A Survey on Explainable Artificial Intelligence (XAI)." *IEEE Access* 6: 52138–52160.
Akaike, Hirotsugu. 1974. "A New Look at the Statistical Model Identification." *IEEE Transactions on Automatic Control* 19 (6): 716–723.
Anderson, David R. 2008. *Model Based Inference in the Life Sciences: A Primer on Evidence*. Springer.
Bandyopadhyay, Prasanta S., and Malcolm Forster, eds. 2010. *Philosophy of Statistics*. Handbook of the Philosophy of Science. Elsevier.
Battaly, Heather. 2008. "Virtue Epistemology." *Philosophy Compass* 3 (4): 639–663.
Berger, James O., and Robert L. Wolpert. 1988. *The Likelihood Principle*. 2nd ed. Institute of Mathematical Statistics.
Bianchini, Stefano, Moritz Müller, and Pierre Pelletier. 2020. "Deep Learning in Science." *arXiv:2009.01575* [cs.CY].
Bickel, P.J., E.A. Hammel, and J.W. O'Connell. 1975. "Sex Bias in Graduate Admissions: Data from Berkeley." *Science* 187 (4175): 398–404.
Bird, Alexander, and Emma Tobin. 2018. "Natural Kinds." In *The Stanford Encyclopedia of Philosophy*, Spring 2018, edited by Edward N. Zalta. Meta-physics Research Lab, Stanford University Press.
Birnbaum, Allan. 1962. "On the Foundations of Statistical Inference." *Journal of the American Statistical Association* 57 (298): 269.
Bonjour, Laurence. 1980. "Externalist Theories of Empirical Knowledge." *Midwest Studies in Philosophy* 5 (1): 54–74.
Box, George E.P., Alberto Luceño, and Maria del Carmen Paniagua-Quinones. 2009. *Statistical Control by Monitoring and Adjustment*. Wiley.
Boyd, Richard. 1999. "Homeostasis, Species, and Higher Taxa." In *Species: New Interdisciplinary Essays*, edited by Robert A. Wilson, 141–186. MIT Press.
Bradley, Darren. 2015. *A Critical Introduction to Formal Epistemology*. Bloomsbury.
Cartwright, Nancy. 1983. *How the Laws of Physics Lie*. Oxford University Press.

———. 1999. *The Dappled World*. Cambridge University Press.
Childers, Timothy. 2013. *Philosophy and Probability*. Oxford University Press.
Conee, Earl, and Richard, Feldman. 1998. "The Generality Problem for Reliabilism." *Philosophical Studies* 89 (1): 1–29.
Dennett, Daniel C. 1991. "Real Patterns." *The Journal of Philosophy* 88 (1): 27–51.
Earman, John. 1992. *Bayes or Bust? A Critical Examination of Bayesian Confirmation Theory*. The MIT Press.
Edwards, Ward, Harold Lindman, and Leonard J. Savage. 1963. "Bayesian Statistical Inference for Psychological Research." *Psychological Review* 70 (3): 193–242.
Efron, Bradley, and Trevor Hastie. 2016. *Computer Age Statistical Inference: Algorithms, Evidence, and Data Science*. Cambridge University Press.
Forster, Malcom, and Elliott Sober. 1994. "How to Tell when Simpler, More Unified, or Less Ad Hoc Theories Will Provide More Accurate Predictions." *The British Journal for the Philosophy of Science* 45 (1): 1–35.
Fumerton, Richard. 2002. "Theories of Justification." In *The Oxford Handbook of Epistemology*, edited by Paul K. Moser, 204–233. Oxford University Press.
Galton, Francis. 1886. "Regression Towards Mediocrity in Hereditary Stature." *The Journal of the Anthropological Institute of Great Britain and Ireland* 15: 246–263.
Gelman, Andrew, John Carlin, Hal Stern, and Donald Rubin. 2004. *Bayesian Data Analysis*. Chapman & Hall/CRC.
Gelman, Andrew, and Cosma Rohilla Shalizi. 2012. "Philosophy and the Practice of Bayesian Statistics." *British Journal of Mathematical and Statistical Psychology* 66 (1): 8–38.
Gettier, Edmund L. 1963. "Is Justified True Belief Knowledge?" *Analysis* 23 (6): 121–123.
Gillard, Jonathan. 2020. *A First Course in Statistical Inference*. Springer.
Gillies, Donald. 2000. *Philosophical Theories of Probability*. 1st ed. Routledge.
———. 2009. "Problem-Solving and the Problem of Induction." In *Rethinking Popper*, edited by Zuzana Parusniková and Robert S. Cohen, 103–115. Springer.
Glymour, Clark. 1981. *Theory and Evidence*. Chicago University Press.
Glymour, Clark, and Madelyn R. Glymour. 2014. "Commentary: Race and Sex Are Causes." *Epidemiology* 25 (4): 488–490.
Goldman, Alvin I. 1975. "Innate Knowledge." In *Innate Ideas*, edited by Stephen P. Stich, 111–120. University of California Press.
———. 2009. "Internalism, Externalism, and the Architecture of Justification." *The Journal of Philosophy* 106 (6): 309–338.
Goodfellow, Ian, Yoshua Bengio, and Aaron Courville. 2016. *Deep Learning*. The MIT Press.
Hacking, Ian. 1980. "The Theory of Probable Inference: Neyman Peirce and Braithwaite." In *Science, Belief and Behavior: Essays in Honour of R. B. Braithwaite*, edited by D. H. Mellor, 141–160.
———. 1990. *The Taming of Chance*. Cambridge University Press.
———. 2006. *The Emergence of Probability: A Philosophical Study of Early Ideas about Probability, Induction and Statistical Inference*. Cambridge University Press.
———. 2016. *Logic of Statistical Inference*. Cambridge University Press.
Hájek, Alan. 2007. "The Reference Class Problem is Your Problem Too." *Synthese* 156 (3): 563–585.
Hand, David J. 2008. *Statistics: A Very Short Introduction*. Oxford University Press.
Hasan, Ali, and Richard Fumerton. 2018. "Foundationalist Theories of Epistemic Justification." In *The Stanford Encyclopedia of Philosophy*, Fall 2018, edited by Edward N. Zalta. Metaphysics Research Lab, Stanford University.

Hempel, Carl G., and Paul Oppenheim. 1948. "Studies in the Logic of Explanation." *Philosophy of Science* 15 (2): 135–175.

Hendricks, Lisa Anne, Zeynep Akata, Marcus Rohrbach, Jeff Donahue, Bernt Schiele, and Trevor Darrell. 2016. "Generating Visual Explanations." *Computer Vision—ECCV 2016*. ECCV 2016. Lecture Notes in Computer Science 9908.

Hoff, Peter D. 2009. *A First Course in Bayesian Statistical Methods*. Springer.

Holland, Paul W. 1986. "Statistics and Causal Inference." *Journal of the American Statistical Association* 81 (396): 945–960.

Howson, Colin, and Peter Urbach. 2006. *Scientific Reasoning: The Bayesian Approach*. 3rd ed. Open Court.

Hume, David. 1748. *An Enquiry Concerning Human Understanding*.

Imaizumi, Masaaki, and Kenji Fukumizu. 2019. "Deep Neural Networks Learn Non-Smooth Functions Effectively." In *Proceedings of Machine Learning Research*, edited by Kamalika Chaudhuri and Masashi Sugiyama, 89: 869–878.

Iten, Raban, Tony Metger, Henrik Wilming, Lidia del Rio, and Renato Renner. 2020. "Discovering Physical Concepts with Neural Networks." *Physical Review Letters* 124 (1): 010508.

James, William. 1907. *Pragmatism: A New Name for Some Old Ways of Thinking*. Harvard University Press.

Jaynes, E. T. 1957. "Information Theory and Statistical Mechanics." *Physical Review* 106 (4): 620–630.

Jeffrey, Richard. 2004. *Subjective Probability: The Real Thing*. Cambridge University Press.

Kasuya, Eiiti. 2015. "Misuse of Akaike Information Criterion in Ecology—AIC Does Not Choose Correct Models Because AIC Is Not Aimed at the Identification of Correct Models (in Japanese)." *Japanese Journal of Ecology* 65 (2): 179–185.

Kelleher, John D. 2019. *Deep Learning*. The MIT Press.

Klein, Peter D. 1999. "Human Knowledge and the Infinite Regress of Reasons." *Philosophical Perspectives* 13: 297–325.

Konishi, Sadanori, and Genshiro Kitagawa. 2008. *Information Criteria and Statistical Modeling*. Springer Science & Business Media.

Krantz, David H., Patrick Suppes, R. Duncan Luce, and A. Tversky. 1971. *Foundations of Measurement (Additive and Polynomial Representations)*. Vol. 1. Academic Press, Inc.

Krohn, Jon, Grant Beyleveld, and Aglaé Bassens. 2019. *Deep Learning Illustrated: A Visual, Interactive Guide to Artificial Intelligence*. Addison-Wesley Professional.

Kutach, Douglas. 2014. *Causation*. Polity Press Ltd.

Leemis, Lawrence M., and Jacquelyn T. McQueston. 2008. "Univariate Distribution Relationships." *American Statistician* 62 (1): 45–53.

Leeow, Jan de. 1992. "Introduction to Akaike (1973) Information Theory and an Extension of the Maximum Likelihood Principle." In *Breakthroughs in Statistics*, edited by Kotz S. and N. L. Johnson, 599–609. Springer Series in Statistics. Springer.

Lewis, David. 1973. "Causation." *The Journal of Philosophy* 70 (17): 556–567.

———. 1980. "A Subjectivist's Guide to Objective Chance." In *Studies in Inductive Logic and Probability*, edited by Richard C. Jeffrey, II:263–293. University of California Press.

Lindley, Dennis V., and Lawrence D. Phillips. 1976. "Inference for a Bernoulli Process (a Bayesian View)." *American Statistical* 30 (3): 112–119.

Marcellesi, Alexandre. 2013. "Is Race a Cause?" *Philosophy of Science* 80 (5): 650–659.

Mayo, Deborah G. 1996. *Error and the Growth of Experimental Knowledge*, 493. University of Chicago Press.

———. 2018. *Statistical Inference as Severe Testing: How to Get Beyond the Statistics Wars.* Cambridge University Press.

McGrayne, Sharon Bertsch. 2011. *The Theory that Would Not Die: How Bayes' Rule Cracked the Enigma Code, Hunted Down Russian Submarines, & Emerged Triumphant from Two Centuries of Controversy.* Yale University Press.

Millikan, Ruth Garrett. 1984. *Language, Thought, and Other Biological Categories.* The MIT Press.

Morgan, Stephen L., and Christopher Winship. 2007. *Counterfactuals and Causal Inference.* Cambridge University Press.

Mumford, Stephen, and Rani Lill Anjum. 2014. *Causation: A Very Short Introduction.* Oxford University Press.

Nagel, Jennifer. 2014. *Knowledge: A Very Short Introduction.* Oxford University Press.

Narens, Louis. 2007. *Introduction to the Theories of Measurement and Meaningfulness and the Use of Symmetry in Science.* Psychology Press.

Neyman, Jerzy. 1957. "'Inductive Behavior' as a Basic Concept of Philosophy of Science." *Revue de l'Institut International de Statistique/Review of the International Statistical Institute* 25 (1/3): 7.

Neyman, Jerzy, and Egon S. Pearson. 1933. "On the Problem of the Most Efficient Tests of Statistical Hypotheses." *Philosophical Transactions of the Royal Society A: Mathematical, Physical and Engineering Sciences* 231 (694–706): 289–337.

Nozick, Robert. 1981. *Philosophical Explanations.* Harvard University Press.

Otsuka, Jun. 2019. "Design Problems in Life and AI (in Japanese)." *Journal of the Japan Association for Philosophy of Science* 46 (2): 71–77.

Pappas, George. 2017. "Internalist vs. Externalist Conceptions of Epistemic Justification." In *The Stanford Encyclopedia of Philosophy*, Fall 2017, edited by Edward N. Zalta. Metaphysics Research Lab, Stanford University.

Pearl, Judea. 2000. *Causality: Models, Reasoning, and Inference.* Cambridge University Press.

———. 2009. *Myth, Confusion, and Science in Causal Analysis.* Technical Report.

Pearl, Judea, Madelyn Glymour, and Nicholas P. Jewell. 2016. *Causal Inference in Statistics: A Primer.* John Wiley & Sons.

Pearson, Karl. 1892. *The Grammar of Science.* Adam/Charles Black.

Peters, Jonas, Dominik Janzing, and Bernhard Schölkopf. 2017. *Elements of Causal Inference: Foundations and Learning Algorithms.* The MIT Press.

Ponciano, José Miguel, and Mark L. Taper. 2019. "Model Projections in Model Space: A Geometric Interpretation of the AIC Allows Estimating the Distance Between Truth and Approximating Models." *Frontiers in Ecology and Evolution* 7.

Popper, Karl Raimund. 1959. *The Logic of Scientific Discovery*, 513. Hutchinson & Co.

Porter, Theodore M. 1996. *Trust in Numbers*, 324. Princeton University Press.

———. 2001. "Statistical Tales." *American Scientist* 89 (5): 469–470.

Pritchard, Duncan. 2014. *What Is This Thing Called Knowledge?* 3rd ed. Routledge.

Quine, Willard Van Orman. 1951. "Two Dogmas of Empiricism." *The Philosophical Review* 60 (1): 20–43.

———. 1960. *Word and Object.* The MIT Press.

———. 1986. "Reply to Morton White." In *The Philosophy of W. V. Quine*, edited by L. E. Harn and P. A. Schipp. Open Court.

Reichenbach, Hans. 2008. *The Concept of Probability in the Mathematical Representation of Reality.* Open Court.

Ribeiro, Marco Tulio, Sameer Singh, and Carlos Guestrin. 2016. "Why Should I Trust You?: Explaining the Predictions of Any Classifier." *KDD '16: Proceedings of the 22nd*

ACM SIGKDD International Conference on Knowledge Discovery and Data Mining, 1135–1144.
Rorty, Richard. 1979. *Philosophy and the Mirror of Nature*. Princeton University Press.
Rosenbaum, Paul. 2017. *Observation and Experiment: An Introduction to Causal Inference*. Harvard University Press.
Roush, Sherrilyn. 2005. *Tracking Truth: Knowledge, Evidence, and Science*. Oxford University Press.
Rowbottom, Darrell P. 2015. *Probability*. Polity Press.
Rubin, Donald B. 1974. "Estimating Causal Effects of Treatments in Randomized and Nonrandomized Studies." *Journal of Educational Psychology* 66 (5): 688–701.
Russell, Bertrand. 1948. *Human Knowledge: Its Scope and Limits*. Allen & Unwin.
Salsburg, David. 2001. *The Lady Tasting Tea: How Statistics Revolutionized Science in the Twentieth Century*. Macmillan.
Sellars, Wilfrid. 1997. *Empiricism and the Philosophy of Mind*. Harvard University Press.
Shibamura, Ryo. 2004. *R.A. Fisher's Statistical Theory: The Formation and Social Context of Inferential Statistics (in Japanese)*. Kyushu University Press.
Simpson, Edward H. 1951. "The Interpretation of Interaction in Contingency Tables." *Journal of the Royal Statistical Society: Series B Statistical Methodology* 13 (2): 238–241.
Sober, Elliott. 1980. "Evolution, Population Thinking, and Essentialism." *Philosophy of Science* 47: 350–383.
———. 2008. *Evidence and Evolution*. Cambridge University Press.
———. 2015. *Ockham's Razors*, 322. Cambridge University Press.
Sosa, Ernest. 1985. "Knowledge and Intellectual Virtue." *Monist* 68 (2): 226–245.
———. 2007. *A Virtue Epistemology: Apt Belief and Reflective Knowledge*. Vol. 1. Oxford University Press.
———. 2009. *Reflective Knowledge: Apt Belief and Reflective Knowledge*. Vol. 2. Oxford University Press.
Spirtes, Peter, Clark Glymour, and Richard Scheines. 1993. *Causation, Prediction, and Search*. The MIT Press.
Staley, Kent, and Aaron Cobb. 2011. "Internalist and Externalist Aspects of Justification in Scientific Inquiry." *Synthese* 182 (3): 475–492.
Steup, Matthias, and Ram Neta. 2020. "Epistemology." In *The Stanford Encyclopedia of Philosophy*, Summer 2020, edited by Edward N. Zalta. Meta-physics Research Lab, Stanford University.
Stich, Stephen P. 1990. *The Fragmentation of Reason: Preface to a Pragmatic Theory of Cognitive Evaluation*. The MIT Press.
Stigler, Stephen M. 1986. *The History of Statistics: The Measurement of Uncertainty Before 1900*. Harvard University Press.
Stone, Mervyn. 1977. "An Asymptotic Equivalence of Choice of Model by Cross-validation and Akaike's Criterion." *Journal of the Royal Statistical Society* 39 (1): 44–47.
Suárez, Mauricio. 2020. "Philosophy of Probability and Statistical Modelling." In *Elements in the Philosophy of Science*. Cambridge University Press.
Szegedy, Christian, Wojciech Zaremba, Ilya Sutskever, Joan Bruna, Dumitru Erhan, Ian Goodfellow, and Rob Fergus. 2014. "Intriguing Properties of Neural Networks." *arXiv, 1312.6199*.
Takeuchi, Kei. 1976. "Distribution of Information Statistics and Criteria for Adequacy of Models (in Japanese)." *Suri-Kagaku (Mathematical Science)* 153: 12–18.
Tanaka, Akinori, Akio Tomiya, and Koji Hashimoto. 2021. *Deep Learning and Physics*. Springer.

VanderWeele, Tyler J., and Whitney R. Robinson. 2014. "On the Causal Interpretation of Race in Regressions Adjusting for Confounding and Mediating Variables." *Epidemiology* 25 (4): 473–484.

van Fraassen, Bas C. 1977. "Relative Frequencies." *Synthese* 34: 133–166.

van Fraassen, Bas C. 1980. *The Scientific Image*. Oxford University Press.

Vaughan, Simon. 2013. *Scientific Inference: Learning from Data*. Cambridge University Press.

von Mises, Richard. 1928. *Probability, Statistics, and Truth*. 2nd revised English ed. George Allen/Unwin ltd.

Wasserman, Larry. 2004. *All of Statistics: A Concise Course in Statistical Inference*. Springer.

Wasserstein, Ronald L., and Nicole A. Lazar. 2016. "The ASA's Statement on p-Values: Context, Process, and Purpose." *American Statistical* 70 (2): 129–133.

Williamson, Jon. 2010. *In Defence of Objective Bayesianism*. Oxford University Press.

Wimsatt, William C. 2007. *Re-Engineering Philosophy for Limited Beings: Piecewise Approximations to Reality*. Harvard University Press.

Woodward, James. 2003. *Making Things Happen*. Oxford University Press.

Worrall, John. 2007. "Why There's No Cause to Randomize." *The British Journal for the Philosophy of Science* 58 (3): 451–488.

Xiao, Kai, Logan Engstrom, Andrew Ilyas, and Aleksander Madry. 2020. "Noise or Signal: The Role of Image Backgrounds in Object Recognition." *arXiv: 2006.09994* [cs.CV].

Zagzebski, Linda Trinkaus. 1996. *Virtues of the Mind: An Inquiry into the Nature of Virtue and the Ethical Foundations of Knowledge*. Cambridge University Press.

INDEX

σ-algebra 19, 29, 39nn5–6

activation function 125, 128
adversarial example 137–138, 142, 143n11, 176
affirming the consequent 79
AIC 116–124, 128n3, 129, 142, 179
Akaike, H. 114–115, 117–118, 124
Alpha Go 133, 135, 139
alternative hypothesis 80–85, 91–92, 95, 99, 101, 107n6, 151
ampliative inference 53
a posteriori 53, 62–63, 67, 71, 133, 181
a priori 53, 61–64, 67, 133, 152, 180–181
Arbuthnot, J. 74
Aristotle 4, 131, 135
artificial intelligence (AI) 129, 141; explainable (XAI) 138, 141
atomism 14
attention 137
average 12–13, 116–118, 177; *see also* mean
average treatment effect 154–159, 164, 166, 169, 174

back door criterion 164
backpropagation method 127
base rate fallacy 59, 107n5
Bayes: empirical 63–66; objective 73n9
Bayesianism 41, 51, 75, 98–106, 107n8, 107n12, 122, 129, 171n3, 176, 179–181
Bayesian network 170
Bayes' theorem 21, 46–47, 49–50, 52–53, 56, 58, 69–70, 98, 104

belief 54–58, 65–71, 88–92, 122, 130, 134–135, 138; basic 61–66, 69; degree of 42–47, 56–58, 62–65, 70, 73n10, 78, 107n8, 104, 176; network 68–71; *see also* justified true belief
benchmark 133, 135, 141, 181
Bertrand paradox 63
biology 4, 35, 40n15, 42, 132, 139, 143n8
Boltzmann, L. 14
Boolean algebra 43
Box, G. 29, 40n13, 131, 142

Cartwright, N. 40n13, 119, 171n6
causal dependence 149
causal discovery 165–166, 171
causal graph 147, 159–166, 168, 170, 173, 178
causal inference 144, 154–159, 162, 166–169, 173–174; the fundamental problem of 153, 156, 179; and tests 151–153
causality 2, 8, 15, 144–152, 158–159, 162–164, 167–169, 178; counterfactual theory of 149–159, 164, 171, 178; regularity theory of 145–149, 158, 171n1
causal kind 170
causal Markov condition 8, 161–167, 171n6
causal model 158–162, 167–169, 171, 173, 178
cause 15, 145, 149, 151–154, 171n3, 178; common 145–146, 160, 165–166
central limit theorem 26–27, 32, 40n11
chemistry 4, 35–38, 40n15, 119, 132, 175

Index

coherence 13, 53, 58, 64, 70, 94, 106, 179–180
collective 75, 77–78, 86, 107n4, 177
collider 160–161, 165–166
confidence 84–86, 91–92, 95–96, 103, 107n10, 133, 151; interval 114
confounding factor 146–148, 155–159, 162, 165
consistency 44, 52–53, 57, 77, 101, 104, 140; statistical 122
constant conjunction 15–16, 145–147
contradiction 43, 72n1, 73n6, 102
convergence 26–28, 34, 60–61, 70, 75, 122, 141, 143n3
correlation coefficient 12, 15–16, 32–33, 170
counterfactual 90–94, 104, 107n12, 108n14, 142n1, 148–149, 156; conditional 90–96, 103, 149–152; model 157–159, 162, 164, 167, 171, 178
covariate 143n5, 146–148, 157–159, 164–165
covering law model 167
critical region 81–86, 99
cycle 160

deduction 52–53, 56–58, 71
deep learning 7, 124–129, 131–143, 176, 180–182
deep learning model 124–142, 176, 181; branding of 133; *see also* neural network
Dennett, D. 120, 175
dependence 20, 145–147, 158, 161, 166
Descartes, R. 2, 61, 181
descriptive statistics 10–18, 33, 38, 112, 172
discrete variable 13, 23–24, 48, 143n4
distribution 21–29, 34, 37, 39n2, 40nn11–13, 46, 96, 107nn4–10, 110, 113–114, 120, 122, 129, 165–170, 171n9, 174; Bernoulli 30–32, 34, 36, 46–50, 60, 174; beta 33, 50; binomial 30–32, 34, 36–37, 49, 81–84, 96, 110–111; exponential 33, 50; family of 29–38, 40n12, 66–67, 93, 118, 174; multinomial 36; multivariate normal 32; normal 27, 30–34, 36–37, 50, 62, 67, 113–114, 174; Poisson 33, 50; uniform 30, 48–49, 62
do-calculus 163–164, 168
double dipping 66
dropout 128
d-separation 161, 165–166, 169
Dutch book 45, 63

Earman, J. 59, 61, 72
economy of thought 13–16, 172–173
empiricism 16, 38
engineering 7, 136–137, 180
entropy 73n9
enumerative induction 34, 40n14
episteme 2, 54, 131
epistemic agent 43–46, 50, 56–64, 70–71, 73n10, 76–77, 88, 90, 94–97, 134–136, 177–178, 181
epistemic virtue 131, 134–137, 181
epistemology 2–7, 38, 43, 53–56, 101–106, 130, 151, 165, 179–182; externalist 55, 87–98, 103–106, 133; holism 68–71; internalist 56–60, 65–66, 69–72, 88–90, 104–106, 180; pragmatist 96, 129–132, 142, 180; virtue 7, 134–136, 142, 181
error function *see* loss function
error term 113–114, 146, 162
estimation 18, 33, 71, 117–120, 142, 156–157, 169–170, 174, 182
estimator 26, 39n1, 42, 111–112, 117–118, 122
ethics 134–135
event 15–6, 18–23, 39n5, 42–43, 64, 74–77, 145, 149, 176–178
evolution 133, 135, 143nn8–9
exclusive 19–20, 76, 171n5
exogenous variable 162
expected value 24–26, 28, 33, 39n1, 40n12, 164, 169, 174
experimental design 68–71, 98–104
explanatory variable 113–115, 118, 124, 143n5, 146–147
externalism *see* epistemology, externalist

factorization 163–164
fair odds 44
faithfulness 166–167, 169, 174
fire alarm 152
Fisher, R. 74, 155
fitted model 112, 116–117, 121
fitting 112, 115–117, 126–127
foundationalism 61–62, 65, 69, 133, 181
frequency 42, 64, 75–78, 106n2, 107n3, 176
frequentism 3, 42–43, 75–78, 85–86, 93–108, 110, 114, 176–178

Galileo 181
Galton, F. 13–15, 17, 112
gavagai 140–141
Gelman, A. 67–69, 71
generality problem 101–103

generalization capability 128–129, 140, 180
generalized linear model 167
generative adversarial networks (GAN) 133
Gettier problem 88–89, 91, 107n11
Gillies, D. 44, 62–63, 72–73, 77, 106–107
given 65, 71, 103, 179
Glymour, C. 39n3, 73n8, 159, 162, 171, 178
Goldbach's conjecture 55
Goldman, A. 58, 89
GPT-3 133
graph 22, 110–111; directed 8, 160, 162, 170; directed acyclic (DAG) 160; *see also* causal graph

habit of mind 17, 87
homomorphism 177
Howson, C. 52–53, 72, 99, 101
Hume, D. 2, 4, 15–18, 25–26, 28–29, 87, 105, 142, 144–147, 171n1
hypothesis testing 33, 78, 90, 112, 151, 171n9; *see also* statistical test
hypothetico-deductive method 67, 71, 79

ignorability 8, 157–159, 164–166, 169, 174
IID 25–28, 39nn9–10, 68, 73n5, 98, 166, 174
independence 20, 23, 25–26, 145, 154–157, 161, 166–167, 169–170, 171n9; conditional 146, 154, 157, 166, 170
independent variable 143
indeterminacy of translation 140
induction 4, 16–17, 34, 37, 40n14, 87, 109, 144–145
inductive behavior 85–88
inductive logic 51–53
inferential statistics 4–6, 17–18, 25–29, 33–38, 42, 74, 105, 114, 118, 121, 129, 167–173
infinitism 73n7
internalism *see* epistemology, internalist
intersubjectivity 62–63
intervention 162–165, 168–170, 171n8, 173, 178; fat-hand 171n8

jade 175
James, W. 122
justification 5, 54–58, 73n6, 88–91, 101–106, 107n11, 130–135, 141, 167, 179–181; Bayesian 57–71; in deep learning 133–134, 137–142; frequentist 91–98, 103
justified true belief 54, 88–89

Kant, I. 2, 177
knowledge 5, 15–16, 53–57, 68, 73n6, 88–91, 95, 107n11, 130, 132–141, 149, 179–181; animal 137–139, 141; reflective 137–139, 141; *see also* justified true belief
Kullback-Leibler divergence 118

large sample theory 26–28
law of large numbers 26–28, 34, 61, 70
law of total probability 20–21, 72n5
learning *see* fitting
least squares method 112, 127
Lewis, D. 63, 71, 149–154, 158
likelihood 33, 46–49, 62, 65, 70–71, 80–85, 98–104, 110–118, 128, 142n1; function 62, 67, 96, 98, 111, 127; justification of 66–69; log 111, 115–119, 122, 127; mean 115–119; principle 66, 98–104, 108n14, 158; *see also* maximum likelihood
Lindley, D.V. 99
local optima 112, 127
logic 43, 52–54, 56–57, 71, 79, 93
logical truth 43–45
long short-term memory (LSTM) 137
loss function 127–131, 180
lottery paradox 73n6
lucky guess 5, 54, 57, 89, 95

machine learning 2, 5, 33, 124, 131–136, 141–142, 175, 180–181
marginalization 20–23, 164
maximum likelihood 110–111, 115, 126, 142n2; estimator (MLE) 111–112, 115–116, 142n2, 143n3; method 109–112, 115, 126–127; *see also* likelihood
Mayo, D. 93, 96, 102, 105–106
mean 13, 31–33, 113–119; population 24–27, 40n11, 50, 154–155; sample 11, 13, 25–27, 159
metaphysics 2, 15–16, 69, 130, 135, 140, 142, 153, 156, 158, 179
mirror of nature 130
missing 154, 156
model check 67–71, 73n12
model selection 7, 38, 41, 59, 66, 109, 112–124, 175, 180
modus tollens 79
M-structure 147–148, 158–159, 165
multilayer perceptron *125*
multiple testing 97, 103

natural kind 35–39, 40n15, 40n17, 66, 96, 119–122, 139–142, 174–176
neural network: convolutional 128, 137; deep 124–128, 138, 176; recurrent(RNN) 137
Neurath, O. 68–70, 73n13
Neyman, J. 74, 80, 86–88, 92, 107n9
non-informative prior 61–63
non-parametric statistics 29
normality test 67
Nozick, R. 7, 90–94, 103, 106, 107n12, 149–152
null hypothesis 80–88, 91–103, 110, 146, 151–152

ontology 4–6, 8, 15–18, 33–38, 41–42, 105, 119–124, 144–146, 159, 165, 167–170, 172–176; dualism 18, 167, 173; layers of 168–169, 174–178; trialist 168
overfitting 69, 114–115, 127–129

parameter 28–34, 36, 40n12, 41, 46–51, 60, 66, 81, 93–94, 98, 107n6, 107n10, 110–122, 124–129, 142nn1–2, 146, 158, 169, 174–175
parametric statistics 28–30, 33–34, 37, 59, 66, 96, 114, 124, 159
path 159–161, 165–166; blocked 160–161; directed 159, 166
Pearl, J. 147, 159, 162–163, 165, 169, 171
Pearson, E. 80, 92, 107n9
Pearson, K. 15–16, 38, 145
p-hacking 97, 103
phenomena 15–17, 38, 64, 86, 121, 170, 174–180
physics 4, 14, 40n13, 119, 132, 135, 178
Plato 54
Popper, K. 70, 79, 107
population 18, 22–27, 32–33, 169, 179
positivism 13–18, 38–39
possible world 92–93, 149–156, 164, 168–169, 173; semantics 92–93, 149
posterior distribution 48–50, 60–61, 66–68, 70, 101, 110, 114
posterior predictive distribution 51, 67–68, 73n5
posterior probability 6, 46–61, 67, 72n4, 73n6, 98–99, 122, 146, 178
potential outcome 153–158, 164, 170, 178
power 84–86, 91–96, 103, 107, 133, 151
pragmatism 122, 131, 175; see also epistemology, pragmatist

prediction 4–8, 17–18, 50–51, 67–68, 79, 109–122, 124, 129, 131, 162–164, 167–170, 173–176
pre-learning 128
principal principle 63–66, 71, 73n10
principle of conditionality 108n14
principle of indifference 62–63, 71
principle of sufficiency 108n14
prior distribution 48–50, 59–65, 68, 71, 73n9, 110
prior probability 46–47, 50, 58–59, 64, 104; washing out of 58–60
probabilistic kind 36, 41, 48–50, 53, 59, 62, 66–67, 94–96, 104, 110, 113–114, 119–121, 129, 139, 159, 170, 175; see also statistical model
probability: axiom 19–20, 23, 39, 44–46, 53, 63, 76–77; conditional 20–23, 46, 72n5, 107n7, 110, 154, 163–164; density 23–24, 32; function 19–23, 27, 43–44, 52–53, 107n4, 107n7, 137, 145; theory 18–26, 29, 34, 38, 39n2, 53
probability distribution 22–27, 29, 33, 39n2, 50, 72n4, 107, 146, 157, 161–168, 171n9, 173–174; joint 22, 32, 114, 162–163; marginal 23, 162
probability model 17–19, 25–29, 34, 37–40, 41–43, 50, 74–75, 78, 109–110, 118–123, 143n7, 159, 167–169, 172–177, 180
probability of a hypothesis 78, 85–86, 178
propensity score 157–158, 166, 170
proposition 43–46, 52–53, 65, 71–72, 93, 176–177
psychologism 70, 79
p-value 5, 84, 93, 95–98, 103, 106, 179–180; problem 95

Quine, W. V. O. 67–69, 105, 140–141, 180–181

radical translation 140–141
random 17, 77, 116–117
randomized control trial 155–156
random variable 21–24, 30–31, 48–49, 75–76, 107n7, 113, 124, 171n3; continuous 13, 23–24, 30–32, 39n8, 49, 126, 143n4
real pattern 120–124, 140, 175
recipe-like statistics 105
rectified linear function/unit (ReLU) 125, 134

reference class 64, 102
regress 58–61, 65, 70–73; problem 58
regression: analysis 13–14, 145–147, 158, 170; coefficient 12–13, 114, 125, 147; line 13–17, 113; linear 113–114, 120, 123, 125–126, 146; model 112–114, 124–126, 146–147, 147, 157, 174; toward the mean 13
regularization 69, 129
reliabilism 7, 88–91, 102, 132–133
reliability 7, 86–91, 95–104, 132–134, 151–152, 171n3, 179–181
representation learning 139
representation theorem 177
reproducibility crisis 95–97, 106
response variable 113–114, 118, 124–126, 143n5, 158
Rorty, R. 130
Rubin, D. 157

sample 17–18, 34, 78, 87, 94
sample size 11, 34, 80, 84–85, 122, 143n3, 175
sample space 18–25, 39n2, 42–43, 69, 71, 75–76, 107n4
sample statistics 11–13, 25–27, 33, 169, 172–174
sampling 17–18, 25–28, 51, 118–121, 167–169
satisfiability 52
scale 177
screen-off 72n5, 146
Sellars, W. 65–66, 71
semantics 4–6, 41–43, 64, 69, 75–78, 92–93, 149–151, 158, 172–179
sense data 65
significance 74; level 83–88, 97–101, 107n10
Simpson's paradox 147
skepticism 29, 115, 142
Sober, E. 40n15, 72–73, 80, 86, 99, 122, 142
Socrates 2, 53, 56
Sosa, E. 134–138, 143n9
Spirtes, P. 159, 162
spurious correlation 146–147, 162
standard deviation 11–12

statistical consistency 122
statistical model 27–36, 39n2, 40nn12–13, 66–68, 72n5, 94, 111, 114, 118, 142, 152, 159, 166; *see also* probabilistic kind
statistical test 67–71, 73n12, 74, 78–85, 90–97, 102–103, 146, 151–152, 180
statistics *see* sample statistics
Stich, S. 96, 130–131, 142
stopping rule 99–102, 104
structural causal model *see* causal model
structural equation 162, 174
structural equation modeling (SEM) 167
subjective probability 41–46, 63–65, 69–72, 76–77

techne 131
test *see* statistical test
tracking theory 90–92, 106, 107n12
transfer learning 140
transformer 137
truth-conduciveness 58, 61–62, 66, 70, 94–98, 105, 135–142, 180–181
truth function 52
truth value 52–53, 92–93
type and token 78, 86
type I error 81–85, 101
type II error 81, 83–85

unbiased 39n1, 118, 159
underdetermination 141
uniformity of nature 2, 4, 16–18, 24–29, 34–37, 41, 66–67, 72n5, 115, 120–121, 129, 167–173
universal approximation property 128
Urbach, P. 52–53, 72, 99, 101

validity 52
van Fraassen, B. 106n2, 122
vanishing gradient problem 127–128
variance 31–33, 62, 113–114; population 24–25; sample 12–18
von Mises, R. 77

wine/water paradox 63
Wittgenstein, L. 143n11, 177
Woodward, J. 162

Taylor & Francis eBooks

www.taylorfrancis.com

A single destination for eBooks from Taylor & Francis with increased functionality and an improved user experience to meet the needs of our customers.

90,000+ eBooks of award-winning academic content in Humanities, Social Science, Science, Technology, Engineering, and Medical written by a global network of editors and authors.

TAYLOR & FRANCIS EBOOKS OFFERS:

A streamlined experience for our library customers

A single point of discovery for all of our eBook content

Improved search and discovery of content at both book and chapter level

REQUEST A FREE TRIAL
support@taylorfrancis.com

Printed in Great Britain
by Amazon